PVP

A Critical Review of the Kinetics and
Toxicology of Polyvinylpyrrolidone (Povidone)

Walter Reppe
(1892-1969)

PVP

A Critical Review of the Kinetics and Toxicology of Polyvinylpyrrolidone (Povidone)

B. V. Robinson
F. M. Sullivan

Department of Pharmacology and Toxicology
United Medical and Dental Schools
Guy's Hospital
University of London
London

J. F. Borzelleca

Department of Pharmacology and Toxicology
Medical College of Virginia
Richmond, Virginia

S. L. Schwartz

Department of Pharmacology
Georgetown School of Medicine
Washington, D.C.

CRC Press
Taylor & Francis Group
Boca Raton London New York

CRC Press is an imprint of the
Taylor & Francis Group, an **informa** business

First published 1990 by Lewis Publishers, Inc.

Published 2019 by CRC Press
Taylor & Francis Group
6000 Broken Sound Parkway NW, Suite 300
Boca Raton, FL 33487-2742

First issued in paperback 2019

No claim to original U.S. Government works

ISBN-13: 978-0-367-45083-0 (pbk)
ISBN-13: 978-0-87371-288-0 (hbk)

Visit the Taylor & Francis Web site at
http://www.taylorandfrancis.com

and the CRC Press Web site at
http://www.crcpress.com

Library of Congress Cataloging-in-Publication Data

PVP: a critical review of the kinetics and toxicology of polyvinylpyr-
    rolidone (povidone)
Includes bibliographical references
    1. Povidone—Pharmacokinetics. 2. Povidone—
Toxicology. I. Robinson, B.V. (Brian V.) [DNLM: 1. Povidone—
pharmacokinetics. 2. Povidone—toxicity. WH 450 P994]
RM666.P822P96 1990   615′.7   89-13515
ISBN 0-87371-288-9

Preface

The new era of synthetic polymer chemistry opened by Walter Reppe in the 1930s led to the synthesis of polyvinylpyrrolidone (PVP), which was patented in 1939. The physical, chemical, and physiological properties of this synthetic polymer are the keys to its very widespread use in medicine, pharmaceuticals, cosmetics, foods, printing inks, textiles, and many more diverse applications.

In recent years, the general interest in reassessing the safety of chemicals used in foods, drugs, and cosmetics has led to the reevaluation and reaffirmation of the safety of PVP by the Joint Expert Committee on Food Additives (JECFA) of the World Health Organization (WHO). Along with others, the authors of this book became involved in collecting and reassessing the information, published and unpublished, on the absorption, distribution, storage, and excretion of PVP and in its toxicological evaluation. As 1989 is the 50th anniversary of the invention of PVP, it seemed appropriate to assemble at this time both the published and the unpublished data that have accumulated over the last 50 years, to critically review it, and to assess the safety of PVP as judged by present-day criteria.

In performing this task, extensive literature reviews were carried out, covering the period from 1939 to 1989. The most recent toxicological investigations have been carried out by the two major world manufacturers of PVP, BASF Aktiengesellschaft in Ludwigshafen, Federal Republic of Germany, and GAF Chemicals Corporation in Wayne, New Jersey,

USA, for submission to regulatory authorities and to WHO. Most of this work is unpublished. We have been fortunate in being given access to all of these data and so are able to publish for the first time critiques of much of this invaluable material. The very earliest use of PVP in medicine was during World War II when a 3.5% solution of PVP was infused into patients as a synthetic blood plasma volume expander. This use recognized such valuable properties of PVP as water solubility, viscosity, and osmotic activity and demonstrated that the material appeared to be biologically inert and safe. The toxicity of PVP, extensively studied in a variety of species including humans and other primates, is of extremely low order.

In Chapters 1 and 2 the reader is introduced to PVP, its uses, and its synthesis and properties. We have been assisted in the preparation of this material by both BASF and GAF, who supplied the details of the determination of the K-values of the different PVP polymers and the molecular weight distributions of representative batches of these polymers.

Chapter 3 is a discussion of movement of large molecules across membranes and forms a basis for Chapter 4, which discusses the absorption of PVP following various routes of administration. The excretion and metabolism of PVP are reviewed in Chapter 5, followed by a discussion on the distribution and storage of the material in Chapter 6. Chapter 7 reviews the storage of PVP in humans.

Chapter 8 presents a review of the functional consequences of PVP uptake, with particular emphasis on the reticuloendothelial and immune systems. Chapter 9 is a broad review of the toxicological studies performed on PVP, including acute, subchronic, chronic, reproductive, mutagenicity, and carcinogenicity studies. Each chapter ends with a summary of the main findings. In addition, there is a final, overall summary and conclusions.

We have also prepared an Appendix, listing the key studies, with references, on the absorption, renal elimination, distribution, acute toxicity, subchronic toxicity, chronic toxicity, teratogenicity, mutagenicity, and carcinogenicity of PVP. The Appendix summarizes in list form the major toxico-kinetic and toxicological studies that have been conducted on PVP, giving the species used, the experimental methods employed, the molecular weight of PVP or its K-value, the observations made, and the source of the information.

Acknowledgments

We are grateful to L. Blecher, J. Ansell, and A. S. Wood of GAF Chemicals Corporation, Wayne, New Jersey, USA, and to W. Schwarz and J. Merkle of BASF Aktiengesellschaft, 6700 Ludwigshafen, Federal Republic of Germany, for their assistance, particularly with the chemistry section, and for access to unpublished studies and data.

Contents

List of Figures

List of Tables

List of Tables

1

Introduction

Polyvinylpyrrolidone (PVP, povidone) is the generic name for the water-soluble homopolymer of N-vinyl-2-pyrrolidone (NVP), and in this review it will be referred to simply as PVP. Polyvinylpolypyrrolidone (PVPP, crospovidone) is the generic name of the cross-linked insoluble homopolymer of NVP.

First developed in Germany at I.G. Farben by Professor Walter Reppe and his colleagues during the 1930s, PVP was subsequently widely used as a blood-plasma substitute and extender during World War II. It has the advantages of being nonantigenic, requiring no cross-matching and avoiding the dangers of infectious diseases inherent in blood. In later years, it has been superseded by other materials. However, the issue concerning the effectiveness of 3.5% PVP (K-30) in an isotonic saline solution for parenteral administration was part of the U.S. Drug Efficacy Study Implementation (DESI) review program, and was reported to be effective for the correction of low blood volume in the treatment of shock (*Fed. Reg.*, 1971).

PVP now finds a large number of other uses. It derives its commercial success from its biological compatibility, low toxicity, film-forming and adhesive characteristics, unusual complexing ability, relatively inert behavior toward salts and acids, and its resistance to thermal degradation in solution. Because of these diverse properties, PVP finds important applications in a number of industries, especially the pharmaceutical, food, beverage, cosmetic, toiletry and photo-graphic industries (Hort and Gasman, 1983).

PHARMACEUTICAL USE

PVP, because it becomes sticky when wetted by water and many solvents, is used extensively as a tablet binder. Tablets bound with PVP exhibit reliable rates of drug dissolution, which can also permit its use in sustained-release prepara-tions. PVP is used as a tablet coating aid, and has found use in the formulation of antibiotics, hormones and analgesics for parenteral administration, and also of ophthalmic and topical preparations.

The formulation of poorly water-soluble drugs has chal-lenged the pharmaceutical industry during recent years, and PVP has been found to be useful in the preparation of solids, solutions, and suspensions where enhanced solubility has increased bioavailability. Very recently, it has found use as a carrier to create a gel-like matrix for the transdermal applica-tion of drugs.

Iodine complexed with PVP (povidone iodine) is widely used as a germicide in antimicrobial soaps, surgical hand scrubs, patient preoperative skin cleansers, antiseptics and skin wound cleansers.

FOODS AND BEVERAGES

The WHO Joint Expert Committee on Food Additives (JECFA) has granted an Acceptable Daily Intake (ADI) of 0–50 mg/kg/day for PVP (WHO, 1986). Under United States law PVP has a number of permitted uses in foodstuffs, including use as a binder for vitamin and mineral concen-trate tablets and as a binder for synthetic sweetener tablets. It is also used as a stabilizer for liquid vitamin and mineral concentrates and to prevent crystallization of liquid synthetic sweetener preparations.

It is also used as a diluent and dispersant for food colors. PVP is permitted as a coating for fresh citrus fruit and as a component in the packaging materials which are in contact with aqueous, fatty and dry foods.

PVP has been used in the chill-proofing of beer and the color stabilization of beverages such as white wines, fruit juices and vinegar. PVP has found use because of its ability to complex and form insoluble precipitates with certain polyphenolic compounds, such as the anthocyanogens in beer and wine. PVP has been replaced in this application by the insoluble PVPP.

COSMETICS AND TOILETRIES

PVP and its copolymers are used as a thickener, dispersing agent, lubricant and binder in the cosmetics industry. The polymers are particularly suitable adjuvants for skin cleansing and protection preparations, and hair tints and dressings. Many creams, either greasy or nongreasy, can be prepared from PVP. It acts as a stiffener in hair setting lotions and improves the consistency of shampoos and household detergents to protect the skin. In contrast to anionic colloids, PVP can be combined with cationic substances, e.g., disinfectants or basic dyes, but allowance must be made for the fact that it has a marked affinity for dyes.

PHOTOGRAPHIC PRODUCTS

PVP acts as a protective colloid and silver halide— suspending agent; it is used as a processing aid in the development of silver halide film and to eliminate the occurrence of dichroic stain. As a coating aid, PVP in silver halide emulsions reduces viscosity and increases the covering power of the developed image.

DYEING APPLICATIONS AND INKS

PVP improves the solubility of dye-based inks to give a greater color value per weight of dye. In pigmented inks it increases tinctorial strength, dispersion stability, viscosity and gloss. It is used as a pigment dispersant in paper. It enhances dye receptivity in paper coatings, and hydrophobic fibers such as polyacrylonitrile, polyesters, nylon and cellulosics.

DETERGENTS

PVP functions as a builder in clear, liquid, heavy-duty detergent preparations. It improves soil-suspending capacity.

SUSPENSIONS, DISPERSIONS, AND EMULSIONS

Additions of PVP stabilize aqueous suspensions, dispersions and emulsions. They are adsorbed in a thin layer on the surface of the colloidal particles and thus prevent contact with other colloidal particles.

PRODUCTION OF PLASTICS

In the production of solid polymers, PVP is used as a protective colloid and polymerization stabilizer.

ADHESIVES

In view of its good water solubility, PVP is a useful component in adhesives for rewettable bonding.

PAINTS AND COATINGS

In emulsion paint and other dispersions for coating, PVP finds use as a protective colloid and dispersing agent to influence the rheology of the dispersions.

PAPER AUXILIARIES

In view of its good water solubility, PVP is useful in coated paper that will be rewetted at a later stage. In this application the PVP prevents curling and imparts a smooth surface to the paper.

TABLETTING AUXILIARY

PVP is a binder for all types of tablets for nonpharmaceutical applications, e.g., cleaning tablets for false teeth or paint tablets for children's paint boxes.

OTHER INDUSTRIAL USES

PVP, because of its unique properties, for example, stability over a wide pH range, is useful in various areas of oil recovery; as an additive to cement formulations to increase viscosity and setting time while decreasing fluid loss; in the plastics industry as a particle size regulator, suspending agent and viscosity modifier, and in many other areas of the chemical industry.

2

Synthesis and Properties of PVP

SYNTHESIS OF PVP

PVP is readily prepared by the polymerization of N-vinyl-2-pyrrolidone (N-vinyl-2-pyrrolidinone, NVP). Traditionally, the synthesis of N-vinyl-2-pyrrolidone and the polymerization of the homopolymer have been based on the Reppe acetylene chemistry represented by the following series of reactions:

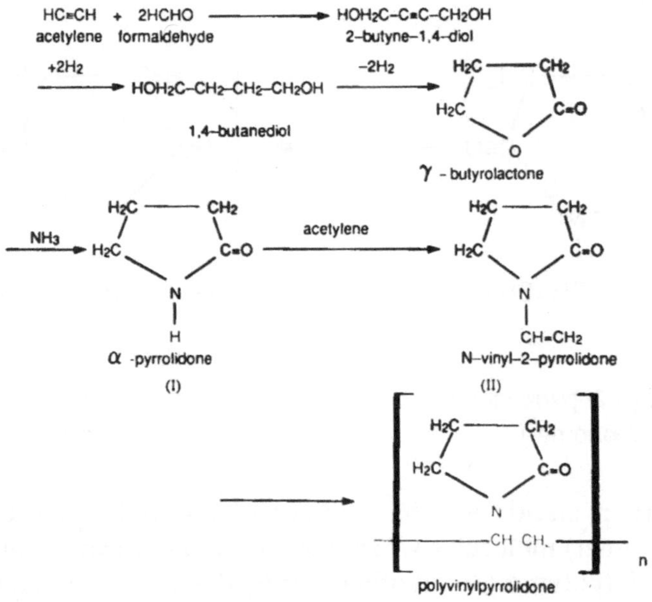

Figure 1. Reppe synthesis of PVP.

The reaction of α-pyrrolidone (I) with acetylene under pressure in the presence of a catalyst yields N-vinyl-2-pyrrolidone (II). A simplified version of polymerization may be visualized as a coupling of the unsaturated $-CH=CH_2$ groups of several N-vinyl-2-pyrrolidone molecules to form a chain of carbon atoms, to which pyrrolidone rings are attached through their nitrogen atoms:

$$-CH-CH_2-CH-CH_2-CH-CH_2-CH-CH_2-$$

The relationship between N-vinyl-2-pyrrolidone (monomer) and PVP (polymer) may be represented more concisely as follows:

N–vinyl–2–pyrrolidone
(monomer)

PVP
(polymer)

where n stands for the number of N-vinyl-2-pyrrolidone (monomer) molecules which have united to form a molecule of PVP (polymer), or in other words, the degree of polymerization of the product.

Currently there are two basic approaches being used commercially for the polymerization of PVP. Most commonly, PVP is polymerized in water as described by Reppe (1949). Used to a lesser extent, but being practiced commercially, is polymerization in isopropyl alcohol.

Both methods yield polymers in solution form. In the isopropyl alcohol method the alcoholic solution is subsequently converted to an aqueous solution by steam distillation. Since the pharmaceuticals industry prefers powders, the aqueous solution is then dried. Low or medium molecular weight polymers are spray dried but high molecular weight polymers are drum dried.

The cross-linked homopolymer, PVPP, is produced using a caustic catalyst (USP 2,938,017) or by using a cross-linking agent (DP 2,059,484) to form an insoluble hydrophilic polymer. The polymer is insoluble in water and virtually all organic solvents.

NOMENCLATURE

Soluble polymers

Chemical Abstracts Services Registration No: 9003–39–8
Chemical Abstracts Name: 1-ethenyl-2-pyrrolidinone homopolymer

PVP has been known in recent years under a variety of names. Some of these have been used as the "approved names" by various regulatory authorities in different countries. There is still no general agreement, but commonly used names include:

polyvinylpyrrolidone
povidone
polyvidone
polyvidon
polyvidonum
poly (N-vinyl-2-pyrrolidinone)
poly (N-vinylbutyrolactam)

poly (1-vinyl-2-pyrrolidone)
1-vinyl-2-pyrrolidinone polymer
poly[1-(2-oxo-1-pyrrolidinyl)ethylene]

PVP is sold in pharmaceutical grade, i.e., in conformance to the requirements of a number of national pharmacopoeias, and also in other grades with non-pharmaceutical specifications. The pharmaceutical grades are marketed under the trade names Kollidon (BASF) and Plasdone (GAF). The non-pharmaceutical grades are marketed under a variety of names, such as PVP, Peregral ST, Albigen A, and Luviskol.

Insoluble polymers

Polyvinylpolypyrrolidone is also known as:

PVPP
crospovidone

These are sold to the pharmaceutical industry under the trade names Kollidon CL (BASF) and Polyplasdone XL (GAF), and to the beer and wine industries under the trade names Divergan (BASF) and Polyclar (GAF).

CLASSIFICATION OF PVP POLYMERS

PVP is marketed for its various uses at different average molecular weights ranging from 2,500 to 1,200,000 and as described below. The different molecular weight materials are distinguished by a K-number (e.g., K-12, K-15, K-17, K-30 for injectable human and veterinary preparations; K-25 and K-30 for oral pharmaceutical and food use and topical pharmaceutical and cosmetic use; K-90 for oral and topical pharmaceutical use; and all grades are available for industrial use). The properties of the material vary according to the average molecular weight. The polymer molecules are composed of between 12 and 1,350 monomer units (Sanner et al., 1983). These have been shown by light scattering measurements in isotonic solutions to have end to end distances of the coiled polymer in the range of 2.3 nm to 93 nm (Sanner et al., 1983).

DIFFERENT METHODS OF ASSESSING THE MOLECULAR WEIGHT

The molecular weight of a polymer is expressed as an average of the various molecular weights of the units comprising the polymer. Traditionally there are three methods by which average molecular weights are determined:

a) Number average ($\overline{M}_n$) – obtained by laboratory procedures that measure characteristics such as osmotic pressure, depression of freezing point, and elevation of boiling point.
b) Viscosity average ($\overline{M}_v$) – determined by viscosity measurements.
c) Weight average ($\overline{M}_w$) – determined by procedures that measure the size of the molecules, such as light scattering or sedimentation rate measurements.

A molecular weight analysis of PVP K-30 using these techniques might give the following approximate values:

Number average ($\overline{M}_n$)	10,000
Viscosity average ($\overline{M}_v$)	42,000
Weight average ($\overline{M}_w$)	55,000

(A molecular weight of 40,000 has been used most commonly in the literature to define this product).

It is absolutely imperative in any review of the literature to know which molecular weight average is being quoted. The most widely accepted standard, for reasons of technical simplicity and calibration, is the viscosity measurement, and the different molecular weight grades of PVP are commonly distinguished by their K-value. This nomenclature is used for PVP, but few other polymers. The relationship between K-value and viscosity average molecular weight depends on the evaluation of a complex equation (Equation 1) in which various constants must be substituted. In work prior to 1951 the two major manufacturers producing PVP, viz., BASF and GAF, used different constants, and therefore the two companies quoted different viscosity average molecular weights ($\overline{M}_v$) for apparently similar K-values. This nomenclature is accepted by pharmacopoeias and other authoritative bodies worldwide. The K-value is usually determined at 1% wt/vol of a given PVP sample in aqueous solution. The relative viscosity is obtained with an Ostwald-Fenske or Cannon-Fenske capillary viscometer, and the K-value is derived from Fikentscher's equation (Fikentscher and Herrle, 1945).

$$\log \frac{\eta_{rel}}{c} = \frac{75K_o^2}{1 + 1.5\,K_oC} + K_o \tag{1}$$

where $\quad$ K = 1000K_o,
$\quad\quad\quad$ η_{rel} = relative viscosity
$\quad\quad\quad$ c = concentration of the solution in g/100 ml.

Solving directly for K, the Fikentscher equation is converted to:

$$K = [\sqrt{300c \log Z + (c + 1.5\ c \log Z)^2} + 1.5\ c \log Z - c]\ /(0.15c + 0.003c^2)$$

where $\quad$ Z = η_{rel}

The intrinsic viscosity [η] is defined as:

$$[\eta] = \lim_{c \to 0}\left(\frac{\ln \eta_{rel}}{c}\right)$$

and is usually determined by measuring the relative viscosity at a number of concentrations and extrapolating $\ln \eta_{rel}/c$ to zero concentration. It may, however, be approximated from the Fikentscher equation by:

$$[\eta] = 2.303 \, (0.001K + 0.000075 \, K^2)$$

where $[\eta]$ = intrinsic viscosity
$\quad\quad\quad$ K = K-value of sample

Utilizing the Mark-Houwink equation (Scholtan, 1951; Von Hengstenberg and Schuch, 1951), the relationship between intrinsic viscosity and viscosity average molecular weight is given by the equation:

$$[\eta] = k\overline{M}_v^a$$

where   k = the Mark-Houwink constant

The values used by the major manufacturers until recently for use in the above equations were:

$$k = 3.1 \times 10^{-4} \text{ and } a = 0.61 \text{ (BASF)}$$
$$\text{and } k = 1.4 \times 10^{-4} \text{ and } a = 0.7 \text{ (GAF)}$$

It seems likely that the type of molecular weight determination will be altered in the future from viscosity average molecular weight to weight average molecular weight using light scattering techniques (Senak et al., 1987).

In general, throughout the remainder of this book, unless otherwise stated, the viscosity average molecular weight $\overline{M}_v$ is used and abbreviated to MW. However, when reviewing the literature where details of molecular weight determination have not been reported or where K-values have not been given, then the information is simply used as set out in the publication.

A comparison of the viscosity average molecular weight for various K-value grades of PVP has been reported by Bühler and Klodwig (1984), and the values are given in Table 1.

Table 1.   PVP K-Values and Viscosity Average Molecular Weight

K-Value (limits as defined by DAC, 1979)	Viscosity average molecular weight, $\overline{M}_v$
12 (11–14)	3,900 (3,100–5,700)
17 (16–18)	9,200 (7,900–10,800)
25 (24–27)	26,000 (23,000–32,000)
30 (28–32)	42,000 (35,000–51,000)
90 (85–95)	1,100,000 (900,000–1,300,000)

DISTRIBUTION OF MOLECULAR WEIGHTS WITHIN A GIVEN PVP PRODUCT

Typically, for radical type polymerization, no batch of polymer will have the same size molecules. The molecular weight distribution curve of PVP is generally broad due to transfer reactions during the polymerization. These transfer reactions occur more frequently in the preparation of the relatively high molecular weight polymers because the degree of branching is greater than in the lower molecular weight polymers. Synthesized PVP products were reported to have an approximate bell-shaped distribution of different molecular weight compounds (Sanner et al., 1983). This fact was demonstrated using PVP K-15, K-30, K-60 and K-90, which when subjected to gel permeation chromatography yielded data illustrated in the accompanying Figure 2 (Senak et al., 1987).

A study of molecular weight distribution of PVP K-30 was reported by Hort and Gasman (1983) and is reproduced in Table 2. This table also shows the results from gel permeation chromatography of more recent samples of PVP K-30 obtained from BASF and GAF (Schwarz, 1987 – personal communication). As a result of changes in manufacturing practice these more recent samples contain a smaller proportion of low molecular weight material (i.e., molecular weight of 20,000 or less) and a greater proportion with a molecular weight in excess of 40,000. The distribution of molecular sizes from the two companies were very similar.

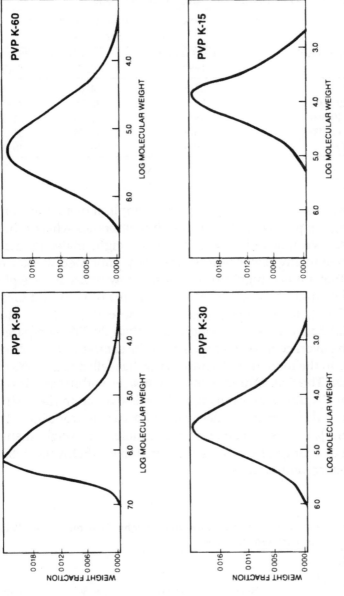

Figure 2.   Gel permeation chromatography of PVP. Absolute differential molecular weight distributions for PVP K-90, K-60, K-30, and K-15 grades.

Table 2.   Analysis of Different Samples of PVP K-30

Molecular weight range	Percentage of different molecular weight ranges present		
	Hort & Gasman (1983)	Schwarz (1987)	
		GAF	BASF
<6000	21.3	4–6	4–6
6,000–20,000	32.2	22–30	25–31
20,000–40,000	19.2	23–25	24–25
40,000–110,000	18.5	30–35	31–37
>110,000	9.0	10–18	10–13

These observations have important implications when considering the passage of PVP through membranes where it has been shown that only low molecular weight materials may penetrate to any real extent. In the early published work it is likely that a much higher proportion of low molecular weight material would be present, and the figures of Hort and Gasman (1983), for example, show that more than 50% of the material has a molecular weight less than 20,000, compared with the average molecular weight of the sample as a whole of 37,900. In the more recent samples this has fallen to 26–37% of the total. The distribution of molecular sizes within samples of a given K-value also means that there is overlap in the distribution of different molecular weights between products with different K-values. An analysis of the molecular weight distribution of a range of PVPs is shown in Table 3 (Schwarz, 1987; personal communication).

Table 3.   Comparison of the Molecular Weight Distribution of PVPs with Different K-values

Molecular weight range by GPC	K-12	K-17	K-30	K-90
< 30,000	99.9	95.0	~47	2–4
30–44,000	0.1	2–3	~12	1–2
> 44,000	–	2.0	~44	96

Table 4. Solubility of PVP in Organic Solvents

Soluble[a]	Insoluble[b]
Alcohols	Hydrocarbons
Methanol	Light petroleum
Ethanol	Toluene
Propanol	Xylene
Butanol	
Acids	Ethers
Formic acid	Diethylether
Acetic acid	
Propionic acid	
Esters	Esters
Ethyl Lactate	Ethyl Acetate
	Sec-Butylacetate
Ketones	Ketones
Methylcyclohexanone	2-butanone
	Acetone
	Cyclohexanone
Chlorinated Hydrocarbons	Chlorinated Hydrocarbons
Methylene Dichloride	Chlorobenzene
Chloroform	Tetrachloromethane
Ethylene Dichloride	
Amines	
Ethylene diamine	
Triethanolamine	
Glycerol	
Glycols	
Diethylene glycol	
Polyethylene glycol 400	
Lactams	
Nitroparaffins	

[a]Minimum of 10% w/w PVP dissolves at room temperature.
[b]Less than 10% w/w PVP dissolves at room temperature.

PHYSICAL AND CHEMICAL PROPERTIES

Physical

PVP is soluble in a variety of organic solvents (Table 4).

PVP is soluble in water, the extent of which for practical purposes is limited only by the viscosity of the resulting solution. The heat of solution is −4.81 kJ/mol (−1.15 kcal/mol); aqueous solutions are slightly acidic (pH 4–5).

Under ordinary conditions, PVP is stable as a solid and in solution. The solid tolerates heating in air for 16 hours at 100°C, but darkening and loss in solubility occurs at 150°C. At pH 12, the polymer gels irreversibly within 4 hours at 100°C. In strong acid solution, PVP is unusually stable, with no change in appearance or viscosity for two months at 24°C in 15% HCl. Studies of various thickening agents for acid gelling showed only PVP to be stable in 15% HCl at 107°C. However, viscosity increases in concentrated hydrochloric acid, and in concentrated nitric acid PVP forms a stable gel.

As shown in Figure 3, the glass transition temperature of PVP is dependent on the molecular weight (Sanner et al.,

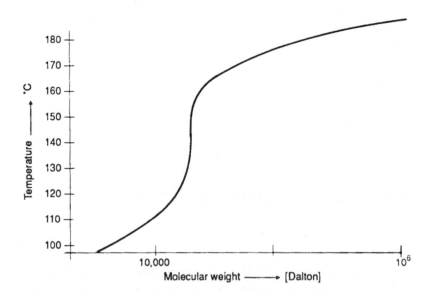

Figure 3. Glass transition temperature vs molecular weight.

1983). The melt viscosity is too high for typical thermoplastic forming operations.

Films of PVP are clear, transparent, glossy and hard. They can be cast from water, methyl alcohol, chloroform or ethylene dichloride. It is comparatively hygroscopic and is between carboxymethyl cellulose (CMC-70) and polyvinyl alcohol (PVA) in absorption of water at 30–90% rh, with CMC-70 > PVP > PVA. At 70% rh, PVP films become tacky, and at 50% rh, they contain 18% moisture.

A number of synthetic and natural resins can be combined with PVP to yield clear solutions and films. Among these compatible resins are ethyl cellulose, polyethylene, polyvinyl chloride, polyvinyl alcohol, polyvinyl methyl ether, shellac, corn dextrin, and polyacrylonitrile (1:3).

Chemical

Polyvinylpyrrolidone is relatively inert to chemical modification. In powder form the various grades of PVP are quite stable at room temperature. When protected from mould, aqueous solutions are also stable. These solutions have no buffering capacity. On standing or by applying heat, they turn slightly yellow, particularly if these solutions are acidic. This discoloration can be inhibited by the addition of reducing agents.

Exposure of PVP solutions to light in the presence of oxidizing agents or diazo compounds can cause gelation (Dorst, 1945; Rose & White, 1945; Slifkin, 1947). When aqueous solutions of PVP are heated with strong bases such as lithium carbonate, trisodium phosphate or sodium metasilicate, a precipitate results. This occurs because of cross-linking, which results from the opening of the pyrrolidone ring and the subsequent reaction across different chains. This precipitation most readily occurs at elevated temperatures (Ullmanns, 1980).

SPECIFICATIONS

The specifications for pharmaceutical grade and typical properties of other grades of PVP are given in Tables 5 and 6.

Table 5.   Specifications of Pharmaceutical Grade of PVP (povidone) (USP, 1990)

Assay	Value
K-Value:	
15 or less	85–115% of nominal value
More than 15	90–108% of nominal value
Water	Not more than 5%
pH, 5% soln.	3.0–7.0
Residue on ignition	Not more than 0.1%
Lead	Limit is 10 ppm
Aldehydes (as acetaldehyde)	Not more than 0.2%
Vinylpyrrolidinone	Not more than 0.2%
Nitrogen	Not less than 11.5% and not more than 12.8%
Hydrazine	Less than 1 ppm

PRESENCE OF CONTAMINANTS IN COMMERCIALLY AVAILABLE PVP

Monomer

PVP is manufactured by polymerization of N-vinyl-2-pyrrolidone (NVP) and, as might be expected, minute quantities of the unchanged monomer are still present in the final product. The USP specification is that this should be less than 0.2%, but in practice the level is normally well below this.

2-Pyrrolidone

PVP also contains small amounts of 2-pyrrolidone, which is due to the hydrolysis of NVP during the polymerization process.

Table 6. Typical Properties of Non-pharmaceutical Grades of PVP

Designation	Form	K-Range	H_2O (% max)	Residual monomer (% max)	Ash (% max)
PVP K-15	Powder	13–19	5.0	0.5	0.02
PVP K-30	Powder	26–34	5.0	0.5	0.02
PVP K-60	Aqueous Soln.	50–62	55	0.5	–
PVP K-90	Aqueous Soln.	80–100	80	0.5	–
PVP K-90	Powder	90–100	5	0.5	0.02

Hydrazine

Commercial PVP contains trace amounts of hydrazine formed from ammonia in the manufacturing process. The limit on the amount of hydrazine permitted in PVP has been set at 1 ppm (USP, 1985; FAO, 1986), but the actual amount present is now much less than this.

Acetaldehyde

Current specifications for pharmaceutical-grade PVP limits the amount of aldehydes to 0.2% as acetaldehyde. It has been reported that nonpharmaceutical grades may contain 0.4% acetaldehyde in PVP K-15, 0.14% in PVP K-30 and 0.014% in PVP K-90 (Ianniello et al., 1987). These values must be regarded with caution, however, since there are major problems in the methodology for the specific determination of acetaldehyde in PVP. New methods are being developed to improve on this methodology (Ianniello, 1989; personal communication).

Based on the very low toxicity observed with PVP, there is no evidence to suggest that these contaminants play any significant role in the toxicity of the final product.

3

Movement of Large Molecules
Across Membranes—
General Considerations

INTRODUCTION

PVP has been widely used medically in the manufacture of tablets and for both diagnostic and therapeutic purposes since the early 1940s. The major use of PVP in quantitative terms has been in the manufacture of tablets. From the amounts of PVP sold for this purpose, it can be calculated that the total number of tablets (including vitamin tablets) manufactured each year is more than 100,000 million (10^{11}) worldwide. From the lack of adverse reactions to this oral administration of PVP over the last 30 or more years, it is reasonable to assume that it is essentially devoid of toxicity in the amounts used. The main parenteral routes of administration have been intravenous as a plasma expander (K-30, average MW 40,000) and subcutaneous or intramuscular as a vehicle for drugs (K-12 to K-17, MW 2,500–10,000).

PVP was consumed in trace amounts in beer, wine and vinegar as a residual contaminant when it was used as a clarifying agent, but PVP is no longer used for this purpose, having been replaced by PVPP. In the United States PVP is used as a stabilizer and dispersant for food dyes and in concentrated liquid sweeteners; and it has been used in coatings on candy, fresh fruit and paper for food wrapping.

Exposure to PVP may also occur by inhalation when it is incorporated as a film former in hair sprays.

The pharmacokinetics of PVP will be discussed under the following headings, covered in this and subsequent chapters:

1. Effect of physicochemical properties of PVP on membrane transport—theoretical considerations
2. Evidence for absorption of PVP following administration orally, subcutaneously, intramuscularly or by inhalation
3. Clearance of PVP from plasma
4. Renal excretion of PVP
5. Metabolism of PVP
6. Biliary excretion of PVP
7. Storage of PVP in the body

Special attention has been paid to the fact that a variety of PVPs with different average molecular weights have been used in these studies and that PVP of a given average MW actually consists of a range of molecules of different sizes. These factors will obviously influence the extent and rates of absorption, distribution, metabolism, excretion, and ultimately storage in the body. The spread of molecular weights within one type of PVP may also lead to difficulties in interpretation of results of studies, especially when only very small proportions of the administered dose appear to be absorbed, as is frequently the case.

PHYSICOCHEMICAL FACTORS

Because PVP is a polydisperse macromolecular polymer that is essentially metabolically inert and can be labelled with radioactive iodine, it has been used to measure pore size of artificial and natural membranes. This has provided a substantial amount of information about the pharmacokinetics of PVP over a wide range of molecular weights. In many instances, the reported average molecular weight of PVP was the basis for estimates of membrane permeability; however, the molecular weight distribution of any given lot of polymers such as PVP reflects a range of molecular sizes in a heterogeneous mixture. For example, a particular lot of PVP K-30 had the molecular weight distribution shown in Table 2.

As discussed earlier in detail, molecular weight measurements also vary according to the methods used to measure and calculate the averages. The number average molecular weight ($\bar{M}_n$), the weight average molecular weight ($\bar{M}_w$) and the viscosity average molecular weight ($\bar{M}_v$) are all different values for PVP. Unfortunately, the method used to determine the molecular weight is not always clearly stated in the literature (perhaps it is not known). The K-value, a reflection of the viscosity of a solution, is a well-established means of expressing relative molecular weight among the various molecular species of PVP. The viscosity average molecular weight has been the most commonly used current expression for PVP. A further discussion of the derivation of viscosity average molecular weight for PVP is included in Chapter 2.

A consideration of the relationship between molecular weight and molecular size is essential to the evaluation of the pharmacokinetics of polymers (including PVP), especially when they are used to determine membrane pore size. The size of a solvated polymeric molecule can be approximated by the unperturbed radius of gyration, S_0, given by:

$$S_0 = \sqrt{\frac{r_0^2}{M}} \times \sqrt{\frac{M}{6}}$$

where r_0 = end to end distance of an unperturbed chain in angstroms

M = molecular weight

$\sqrt{r_0^2}/M$ is a characteristic ratio for a solution of polymers in a defined solvent. Meza and Gargallo (1977) determined this value for a solution of PVP in water to be 0.480 (where r_0 is expressed in angstroms). Therefore:

$$S_0 = 0.480 \sqrt{\frac{M}{6}} = 0.196 \sqrt{M}$$

This equation was used to estimate the radius of a particular molecular weight of PVP in discussions of data in this review.

MOVEMENT OF PVP ACROSS MEMBRANES

Substances may pass across membranes by a number of mechanisms, including simple diffusion, facilitated diffusion, active transport and special mechanisms such as pinocytosis. Most xenobiotics cross membranes by simple diffusion, and for low molecular weight PVPs this is the major mechanism involved. For higher molecular weight PVPs other mechanisms are of importance, as discussed later.

Simple diffusion through a cell membrane occurs by one of two general pathways:

1. Lipophilic molecules diffuse through the lipophilic portion of the membrane.
2. Lipophobic molecules diffuse through the "water pores" in the membrane.

Since the lipophilic portion of the membrane constitutes the largest fraction of the cell surface area, there are few restrictions on molecular size for simple diffusion of lipophilic molecules. Water pores are restricted in size, depending on the site of the membrane. Simple diffusion of lipophobic molecules is subject to limitations based on pore size and molecular size. PVP is lipophobic and its simple diffusion through membranes is subject to such restrictions. Pappenheimer et al. (1951) originally proposed that convection and diffusion act to enhance solute movement through pores, while steric hindrance and friction impede solute movement through the pores. The particular problems relating to the transport of PVP through specific membranes are discussed below.

Glomerular transport of PVP

Much of the data on the movement of PVP through membranes is derived from studies of the glomerular clearance on many different molecular fractions of PVP. Studies by Hardwicke et al. (1968) and Hulme and Hardwicke (1968) provided important information which led to further studies in the refinement of pore transport theories, in general, and the behavior of PVP, in particular. Hulme and Hardwicke (1968) administered radio-iodinated PVP (MW 38,000; range 5,000-80,000) intravenously to normal human subjects. Serial one-hour clearances were measured utilizing serum and urine samples fractionated by gel filtration. Increasing urine/plasma concentration ratios with decreasing molecular sizes were obtained. This relationship continued up to a limiting value, which approximated creatinine urine/plasma concentration ratios in normal individuals.

A series of studies from Lambert's laboratory in Brussels (Gassee et al., 1967; Lambert et al., 1970; Lambert et al., 1971; Lambert et al., 1972; Gassee et al., 1972; Verniory et al., 1973; Verniory et al., 1974; DuBois et al., 1975; DuBois and Stoupel, 1976) significantly expanded our understanding of the permeability of the glomerulus to PVP. Verniory et al. (1973) proposed a transport equation for uncharged macromolecules across a porous membrane which considered the earlier concepts of Pappenheimer et al. (1951). Consider that:

K_1 = function of drag forces on a sphere moving in a stationary liquid

K_2 = function of drag forces on a sphere in a moving liquid

a = molecular radius

r = pore radius

S_D = steric hindrance factor for diffusion; a function of a/r

S_F = steric hindrance factor for convection restricting bulk flow of solute; a function of S_D

D = free diffusion coefficient of the solute in water

n = viscosity of filtrate

Δ_P = filtration pressure

The chemical and physiological factors of concern in pore transport are molecular size, pore radius and filtration pressure. The fractional clearance/GFR is the sieving coefficient, ϕ. As described by Verniory et al. (1973), according to pore theory ϕ is related to a, r and Δ_P by:

$$\phi = [(K_2/K_1)S_F]/1-e^{-k} [1-(K_2/K_1)S_F]$$

where $k = (K_2/D) \cdot (S_F/S_D) \cdot (r^2/8n) \cdot \Delta_P$

DuBois et al. (1975) and Dubois and Stoupel (1976) studied the permeability of artificial porous membranes to PVP. They demonstrated that, considering the influence of filtration pressure on the shape of the sieving curve, the mean pore radius and pore radii distribution calculated from the sieving data correspond to those that would be calculated from the theoretical transport equation for uncharged molecules proposed by Verniory et al. (1973).

The data presented for the glomerular basement membrane and artificial membranes in the studies cited thus far correlate with an assumption that the membranes contain cylindrical pores logarithmically distributed around the median.

The foregoing discussion emphasizes the physical, chemical and biological complexities influencing the movement of PVP through pores and the difficulty in attempting to predict accurately the transport of an average molecular weight PVP across a membrane of average pore size. The failure to recognize these complexities has led to occasional "new theories" to explain the unexpected transport of macromolecules across membranes (including the glomerular membrane) presumed to be impermeable.

4

Absorption of PVP by
Various Routes of Administration

INTRODUCTION

The best method for determining the absorption of a chemical is by direct chemical analysis of the material in the blood and/or tissues. In the case of PVP no satisfactory specific method for the detection of PVP in biological samples has been developed. This has necessitated the use of a variety of other techniques for determining the absorption of PVP. These have included the use of radio-labelled PVP and histological or histochemical investigations. The shortcomings of these techniques are discussed below.

Since PVP is used in pharmaceutical products that are administered orally and in food, there is much more information regarding the oral absorption of PVP than about absorption following administration by other routes. Furthermore, as mentioned earlier, PVP has been used as an experimental tool to investigate the pore size in the gastrointestinal tract and the molecular weight of material that can be absorbed by this route.

ORAL ADMINISTRATION (see Appendix 1)

The experimental evidence regarding uptake of PVP from the gastrointestinal tract comes from three main types of experiments:

1. Observation of gross and histological changes in the tissues, mainly the gastrointestinal tract and associated lymph nodes and the liver, following long-term feeding studies using various grades of PVP. Such observations have generally been made as part of the toxicity studies that have involved feeding large amounts of PVP, generally 1–10% of the total dietary intake of animals.
2. Acute or medium-term studies in which PVP has been administered orally by gavage or given in the food with the specific intention of investigating possible oral absorption. The PVP was estimated in the tissues at post mortem by chemical or histological techniques.
3. Acute experiments in which radio-labelled PVP has been administered orally to conscious animals or by perfusion through the intestine of anesthetized ones. The amount of radioactivity appearing in the urine, blood, feces and tissues or remaining in the perfusate was then estimated to assess absorption. In some studies information about PVP absorption is incidental to the use of this substance as an experimental tool to investigate pore size in membranes.

EVIDENCE OF ABSORPTION FROM TOXICITY STUDIES

A number of subchronic and chronic feeding studies were conducted in the 1950s and early 1960s in rats, cats and dogs using various grades of PVP. These experiments are discussed in detail later in the review in the sections on toxicity and on carcinogenicity. The overall conclusion of the researchers has been that there is evidence only for minimal absorption of PVP, a conclusion that was endorsed by Burnette (1962) when reviewing these and other investigations. All work carried out since that time has confirmed this limited absorption of PVP.

Nine separate rat investigations have been reported, involving over 1200 animals, in which PVP (ranging in size from K-15 to K-90) was administered by gavage (up to 10g/kg) or in the diet (1–10% of diet) over periods ranging from

28 days to 2 years (see Appendices 5 and 6). In none of these studies was there evidence for PVP absorption in terms of either gross or histological changes in tissues (despite, in some instances, special attention being devoted to the lymphatic system for signs of PVP storage in RES cells) or in terms of changes in biochemical parameters in blood or urine where these have been measured.

A similar lack of evidence for PVP absorption was found in cat studies. It was only in the dog that there was possible evidence of absorption and even here the findings have not been consistent. For example, in the short-term feeding studies carried out over 1.5 to 9 weeks, no toxic or pathological changes that could be attributed to PVP were reported (see Appendixes 5A and 5B), although female animals receiving 10% PVP K-90 in a 28-day study did show splenic enlargement (Kirsch et al., 1975). On the other hand, in a 1-year study in which 5% PVP K-30 was administered (see Appendix 6B), and a special staining technique was used to identify PVP in the tissues, positive staining was shown to occur in lymph nodes and blood (but not liver, intestine or spleen) of treated animals. However, since similar findings were also made in the control animals, some doubt must be placed on these observations.

The only unequivocal finding has been in a 2-year dog study (Princiotto et al., 1954) in which 0, 2, 5 or 10% PVP K-30 was administered in the diet to Beagles in combination with 10, 8, 5 or 0% of Solka-Floc, a form of cellulose. At post mortem, although there were no gross or histological changes in a wide variety of organs, it was reported that the reticuloendothelial cells in the lymph nodes were swollen in the group receiving the highest dose of PVP (10%), and that lower-dose groups showed this effect to a lesser extent and less consistently.

Whether this change was due to PVP absorption, and whether the findings are peculiar to dogs, is unclear. They have, however, stimulated a number of more definitive investigations into gastrointestinal absorption.

ABSORPTION STUDIES USING CHEMICAL ANALYSIS OF PVP

Chemical techniques for the identification and quantitation of PVP are more complex and less reliable than those using radio-labelled material. Nevertheless, most of the early studies have, of necessity, used these methods, mainly to assess distribution of intravenously administered PVP where the levels are large and sensitivity is not essential. One study of interest that employed a chemical analytical technique following oral administration of PVP is that of Haranaka (1971), who conducted a number of experiments in rabbits using PVP with a MW of 40,000. In the first he gave the animals 70 g of PVP mixed in the diet over a period of 1 month. After this time cellulose acetate electrophoresis of the plasma revealed slight changes in the protein distribution. The albumin and alpha$_1$-globulin levels were lower than in the controls, while the gamma-globulin level was higher. He attributed this to an effect of PVP on the liver, but conceded that PVP also produced a slight "aggravation in nutrition" which may contribute to the phenomenon. He also showed that PVP could be detected in the liver using the phenol method of detection but not the Lugol method, neither of which is very satisfactory. Correcting for the recovery rate, the highest estimate of liver PVP was less than 30 mg total or 0.04% of the total administered dose.

In a second series of experiments Haranaka perfused 20 ml of 7% PVP (1400 mg) into the upper small intestine of an anesthetized laparotomized rabbit and collected portal blood at intervals over an hour. The serum was separated and PVP estimated using the Lugol method. He found that absorption was maximal after 10 minutes and no further absorption occurred after 30 minutes, but total absorption was only 370 μg, or 0.026% of the administered dose. Haranaka considered this to be an underestimate of the actual absorption for two reasons. First, he showed that as much as two-thirds of the PVP in the blood was bound to red cells and would not be measured by his technique. Second, the site at which he sampled portal blood represented drainage from only part of the intestine, so that the total entering the liver could be higher, although he does not estimate by what proportion. However, even if one takes account of binding to red cells, the amount of PVP absorbed is just over 1 mg. Then, allowing for inadequate sampling (say only 1/5 to 1/10 of the total has been collected) the total absorption would be 5–10 mg, or 0.35–0.7% of the administered PVP.

The overall conclusion drawn by the author was that PVP is absorbed into the circulation and that small amounts were deposited in the liver, probably in the form of a combination with amino acids or peptides at $\geq$CO$=$ and –N in the pyrrolidone radical instead of the original form of PVP. The techniques available to the author clearly did not allow investigation of the molecular size of the absorbed material, but since the PVP that he used (which had a MW of 40,000) in all probability contained more than 20% of molecules with a molecular weight less than 6,000, these could readily account for the PVP found in the body.

ABSORPTION STUDIES USING RADIO-LABELLED PVP

When using radioactive labels to investigate absorption of PVP there are a number of technical problems that can arise which complicate the interpretation of the results. Some relate to the particular isotope used and some to the methods of administration and of measurement.

General problems

First, in all the studies in which molecules with a distribution of molecular sizes have been labelled, it should be remembered that the level of radioactivity in a given sample reflects not the number of molecules present but the weight, because large molecules will have more label attached than small ones. Second, the radio-labelled PVP is, for reasons of safety and economy, diluted with "cold" PVP solution before administration into animals. It is important that both samples should have the same distribution of molecular sizes for accurate assessment of the true rate and extent of absorption of the PVP sample as a whole.

Following on from this, in studies in which PVP is administered orally, it is not sufficient to infer absorption of PVP simply from measurements of radio-label in the collected feces. Loss of radioactivity could simply reflect inadequate collection. Most metabolic balance studies accept recoveries of radioactivity in the region of $100 \pm 5\%$. Absorption of PVP in the region of 5% would therefore be almost undetectable. Low levels of radioactivity in the feces could also be explained by absorption of monomer, free label or only the very low molecular weight material. Conversely, artificially high recoveries may be recorded in the feces because material has been absorbed from the gastrointestinal tract and then subsequently excreted in the bile.

The only real indicator of absorption is the identification of the marker in the blood, tissues and urine. Then, to be absolutely certain that it is PVP and not just free label or monomer, fractionation of the collected radioactivity to identify the molecular weight of the material involved is desirable. Clearly this can only be carried out in urine and tissue fluids, and then only if sufficient material is present. In addition, considerable errors are also likely to arise because of the very low levels of radioactivity absorbed.

Studies using radio-iodine (^{131}I and ^{125}I)

The most important problem in interpreting such studies is that the iodine-label can become unbound, and indeed any sample of such material is likely to contain a minute proportion of free radioactive iodine. In most studies this would be of little consequence, but when considering the potentially very small percentage absorption of PVP through the gastro-intestinal tract it is important to distinguish between the absorption of free radioactive iodine and the absorption of the labelled-PVP material. Few of the many studies using this technique have done so. Dialysis of the starting material reduces the amount of free label, but is unlikely to eliminate it totally.

Studies using radio-carbon (^{14}C)

These studies are undoubtedly the most reliable because the carbon is actually part of the PVP molecule and cannot become unbound. Therefore, unless PVP is metabolized in the body or some of the label is attached to monomer still present in the sample, the distribution of radioactivity will correspond to the distribution of PVP.

Having discussed the technical problems which have to be considered in assessing the results of absorption studies, we will now review the studies which have been carried out.

ABSORPTION STUDIES USING IODINE-LABELLED PVP

One of the earliest investigations was by Fell et al. (1969) using two ewes. They gave the PVP (MW 33,000) directly into the abomasum in one animal and into the duodenum in the other. They showed that peak excretion of radioactivity occurred after about 22 hours, and that during a 120-hour period, 60% and 71% respectively of the administered radio-activity could be obtained from the feces. They assumed that the remaining material had been absorbed, and although radioactivity was detected in the blood throughout the whole collection period, no figures for the extent of absorption were given. Unless the ewe is significantly different from other species, it is unlikely that 30–40% of the adminis-tered PVP had been absorbed. It is possible that the label became detached from the PVP in the gastrointestinal tract, but as the authors pointed out, the lower pH present in the sheep makes this less likely to occur than in other species; furthermore, they presented some evidence from dialysis experiments that this did not occur. To confuse the picture still further, Fell and his colleagues showed that after intravenous injection of ^{131}I-PVP, up to 4.3% of the administered radioactivity was lost in the feces and that this was not due significantly to biliary or pancreatic secretions (as shown in other species). They assumed the PVP entered the feces from the blood through the gastrointestinal wall. One can only conclude that the sheep handles PVP in quite a different manner than other species, or more likely that the study was too small to come to any clear conclusions.

More extensive investigations were conducted by Clarke and Hardy, who in a series of studies (1969a, 1969b, 1971) measured the absorption of ^{125}I-PVP K-60 by weanling rats. They were investigating the uptake of large molecular weight material by the gastrointestinal tract of young rats, and in particular they were interested in protein absorption (e.g., of gamma-globulin). They administered the PVP by stomach tube (5 mg/rat), and 4 hours later, after thorough washing of the intestine, they measured the retention of radioactivity within the gut. They reported that in rats less than 18 days old, up to 50% of the administered PVP K-60 (MW 160,000) was retained. The process of uptake occurred in the distal small intestine (not the duodenum or the large intestine) and was complete within 4 hours. No further uptake took place when longer absorption times were allowed. Uptake declined markedly in rats 18–20 days after birth, until by day 20 less than 5% of the PVP was retained. In a few adult rats tested, uptake of PVP was less than 1%. This change in gastrointestinal function, which is called "closure," paralleled changes in the histological appearance of cells present in the intestine. At about the 18th day, they concluded, there is an abrupt change in the type of cell produced within the crypts of Lieberkuhn, and these have quite different permeability characteristics.

Despite uptake into the intestine, they could not find radioactivity in the blood of the rat, even in those animals in which 50% was taken up by the intestinal wall. Earlier studies by Hardy (1965, 1968) had shown that PVP K-60 was readily transferred from the small intestine into the circulation by the newborn pig and calf. The authors suggest that the uptake of PVP in these species is by pinocytosis, since the level of radioactivity correlates with the degree of vacuolation of intestinal cells. They went on to show that the process is saturable, the uptake being proportional to the administered dose of PVP up to a maximum of only 1 mg. They conclude that this mechanism is probably responsible for the uptake of gamma-globulin from colostrum and milk in young animals, and is an important stage of normal neonatal development. While an uptake in weanling animals of 50% of the administered PVP K-60 must include a large proportion of high molecular weight material, it is not clear to what extent the uptake of radioactivity in adult animals simply reflects the low molecular weight material present in the sample of PVP, or simply free iodine. No attempt was made to fractionate the retained material.

A recent study on the uptake of macromolecules by rat intestine (Beahon & Woodley, 1984) showed that PVP is taken up not only into the columnar epithelium by fluid-phase pinocytosis, but that absorption occurs into Peyer's patches by the same mechanism.

Another study using radio-iodine-labelled PVP was by Loehry et al. (1970) who used a number of substances, including ^{131}I-labelled PVP, to investigate the diameter of pores in the intestine of the rabbit. The experiments were performed in anesthetized animals in which the kidney pedicles were tied off (to prevent urinary loss) and in which the intestine was perfused with physiological saline at 40°C at a rate of 600 ml/hr. Two different procedures were used. First, those in which the radio-labelled PVP was given intravenously and the amount appearing in the perfusate measured, and second, those where the radio-labelled material was recirculated through the intestine and appearance of radioactivity in the blood detected. ^{125}I-labelled PVP with a nominal MW range of 8,000–80,000 with no K-value specified (Radiochemical Centre, Amersham) was used throughout the studies.

In the first series of experiments the "clearance" of substances from the plasma into the intestinal lumen was expressed as follows:

$$\text{Clearance} = \frac{\text{amount of substance excreted/min}}{\text{concentration of that substance in the plasma}}$$

From these studies it was shown, using a variety of agents ranging in molecular weight from urea (MW = 60) to PVP (MW = 33,000), that there is a linear relationship when the logarithm of molecular weight is plotted against the logarithm of intestinal clearance. Intestinal permeability to urea was considered to represent the maximum permeability of the water pores of this tissue.

In addition to these studies measuring total radioactivity, experiments were also conducted in which the plasma and intestinal perfusing fluid were fractionated using Sephadex G200 in order to determine the size of the molecules penetrating through the gastrointestinal mucosa and to confirm that the label was still attached to the PVP molecules of various sizes. This was carried out both in the procedures in which the radio-labelled PVP was infused intravenously and in those in which it was perfused through the intestine. The authors were then able to calculate the relative amounts of ^{125}I-labelled PVP of a given molecular size that had passed from the intestine into plasma or vice versa. Similar results were obtained regardless of the direction of movement of the PVP, but a comparison of the plasma/perfusate ratio of the different molecular weight fractions of PVP essentially showed that as molecular weight increases less material is transported from the intestine into the plasma. The data suggest that PVP with molecular weights of 8,000, 33,000 and 80,000 penetrate the intestine at rates of 0.67%, 0.39% and 0.067% that of urea, respectively. This would indicate that most of the PVP that enters the body is of low molecular weight, with each minute about 10 times as much PVP of molecular weight 8,000 entering as PVP of molecular weight 80,000.

ABSORPTION STUDIES USING ^{14}C-LABELLED PVP

The first study of absorption was carried out by Shelanski (1953) who gave 3.5% ^{14}C-labelled PVP (K-30) solution orally to 5 rats (110–150 g) at a dose of 6–10 g/kg. He claimed that 99% was excreted in the feces over a 5-day period, mostly during the first day, and that about 1% was excreted in the urine, that 0.25% was exhaled as carbon dioxide and that 0.5% of the administered dose remained in the carcass. The study is open to a number of criticisms. First, it was not a balance study and accounted in total for less than 85% of the radioactivity administered. Second, the large dose of PVP which was used caused diarrhea, which made fecal collection unreliable: Shelanski's assumption that 99% of the radioactivity was lost in the feces was based simply on the total activity that could be detected elsewhere (less than 2%). The soft feces may also have led to contamination of the urine samples. Third, although 0.5% of the radioactivity was present in the carcass, less than 0.001% of the radioactivity could be accounted for in the major organs (liver, kidney, lungs, spleen). The source of the remaining activity is unclear, but could represent skin contamination (some attempt was made to wash the animals), or could simply mean that some labelled PVP was still present in the gastrointestinal tract.

A limited but better controlled experiment was conducted by the same author (Shelanski, 1960), but this time using PVP K-90 which was administered orally as a 10% solution at a dose of 5–6 g of solution per rat, equivalent to 500–600 mg of PVP, i.e., 2–2.5 g/kg body weight. In two animals, urine was collected at intervals over a 157-hour period with the aid of a special funnel taped to the penis (to avoid fecal contamination). This showed that a small amount (< 0.35% of the dose) was excreted in the urine over a 13-hour period, with much smaller amounts continuing for at least 5 days. In a third animal, carbon dioxide was collected at intervals over a 168-hour period. This showed that 0.04% of the radioactivity could be detected over the first 6 hours, but after that radioactivity was undetectable. In a fourth animal, feces and urine were collected together and accounted, over a 96-hour period, for 97% of the administered radioactivity. At post mortem, measurement of tissue radioactivity in the animal used in the carbon dioxide collection study showed that, with the exception of the gastrointestinal tract and liver, which contained 0.012% and 0.003% of the administered radioactivity, respectively, levels were not distinguishable from background.

Morgenthaler (1977) carried out more extensive studies in rats, but using PVP K-12 (MW 1,700) and PVP K-25 (MW 40,300) and giving much smaller amounts of PVP (50 mg/kg). Under anesthesia he cannulated the bile duct, the left and right ureters and the jugular vein (to infuse physiological solution, to replace the fluid lost by the animal, and mannitol to create a diuresis). He administered ^{14}C-PVP intraduodenally and collected bile and urine samples at intervals of between 15 and 60 minutes over a 20- to 25-hour period. From six animals treated with PVP K-12, he collected a total of 7.55% of the administered radioactivity in the urine and bile, while in eight animals treated with PVP K-25, only 1.3% was excreted by these routes. Analysis of the entire gastrointestinal tract at the end of the experiment revealed that a mean of 89.6% of the administered K-12 and 93.0% of the K-25 had not been absorbed.

The most definitive ^{14}C-PVP studies to date have been those of Digenis et al. (1987). They used ^{14}C-PVP that had been synthesized by a procedure identical to that used for the commercial production of PVP K-30. This radio-labelled PVP (0.189 g) was dissolved in 20 ml of distilled water to yield a solution containing 106 μCi/ml, and this was administered by gavage (0.1 ml/rat) to each of the test animals. Rats of 200–300 g were used, so that a total dose of only 0.9 mg/rat (about 3–5 mg/kg) was given. A single untreated control group and four test groups, each of five animals, were used.

All the animals were kept in glass metabolism cages which allowed urine and feces to be collected separately. At the end of selected time intervals (6, 12, 24 and 48 hours) the rats were killed and major organs (kidney, stomach, liver, lungs, thymus and spleen) and blood samples were collected for determination of total radioactivity. Because tissue radioactivity in the test animals was very low, background counts were established using tissues from control animals.

Essentially, the study showed that PVP K-30, when administered orally to rats, is absorbed in only trace amounts. By 12 hours the cumulative fecal recovery of radioactivity was 90.8%, and by 48 hours it had risen to 98.4%. Measurement of tissue radioactivity 6 to 48 hours after PVP administration showed only background levels, which were not significantly different from untreated control values. On the other hand, urine collection during the first 6 hours did reveal some radioactivity, with 0.04% of the administered dose being excreted by this route. To investigate further this latter finding, the authors administered ^{14}C PVP orally by gavage to a single rat, which was then anesthetized and the carotid artery cannulated. Hourly blood samples removed over a period of 6 hours revealed that blood radioactivity reached a maximum at 2 hours and that the half-life for disappearance of radioactivity from the blood was 1.5 hours.

Since it was suspected that the material entering the body had a low molecular weight, Digenis et al. (1987) carried out dialysis for up to 309 hours on the sample of ^{14}C PVP used in these studies. At the end of this period they found that only 4.0% of the ^{14}C-PVP had passed through the membrane (i.e., had a molecular weight of less than 3,500). This percentage of low molecular weight material is much less than that found in commercially available PVP K-30. Nevertheless, even this small amount of oligomer would be enough to account for the radioactivity detected in blood and urine in the animal experiments reported above. The authors also showed, using dialysis membrane with a different molecular weight cutoff, that 7.9% of the ^{14}C-PVP had a molecular weight of less than 12,000–14,000. Because of the extremely small amounts of material absorbed and excreted in the urine, it was not possible to determine the molecular weight distribution of the absorbed PVP. Further support for the view that it is the low molecular weight oligomer that is absorbed came from the studies of McClanahan et al. (1984) from the same laboratory, on the disposition of N-vinyl-2-pyrrolidone (NVP, the monomer of PVP). They showed that after intravenous administration of ^{14}C-NVP to the rat, the half-life is similar to that seen with ^{14}C-PVP absorbed from the intestine, namely 1.5 hours. Digenis et al. (1987) concluded that since PVP contains approximately 1% unreacted monomer, this must account for at least some of the absorbed radioactivity.

ABSORPTION STUDIES OF PVP IN HUMANS

While there is much information on humans relating to the effects of injected PVP, there have been only a limited number of studies in which the absorption of orally administered PVP has been investigated, and all of these have been to assess the effects of gastrointestinal disease on permeability. For example, ^{125}I-PVP preparations have been used to measure blood-to-lumen permeability changes in protein-losing gastroenteropathies (Jarnum, 1961; Waldmann, 1972), and lumen-to-blood permeability changes in patients with ulcerative colitis and in experimental models (Rask-Madsen, 1973). However, in all these studies the low levels of absorption of the PVP and the large degree of dissociation of the iodine label meant that they could not be usefully quantified.

Siber et al. (1980) overcame this problem by using ^{14}C-labelled PVP (K-17.8), which they used to monitor the changes in gastrointestinal permeability which occur following the administration of the anticancer drug, 5-fluorouracil, to ten patients with metastatic carcinoma of the colon. Their starting material, although nominally having a molecular weight of 11,000, was shown by Sephadex separation to include molecules with molecular weight in excess of 40,000. To reduce this distribution, they subjected the material to ultrafiltration, using filters to exclude molecules greater than 50,000 daltons and also those less than 20,000 daltons. This removed in total about 30% of the original material. The remaining PVP was administered orally to fasted patients at weekly intervals. Urine and fecal collections were made at 24 hours, and the recovered radioactivity was expressed as a percentage of the total administered.

The results indicated that total fecal excretion of PVP was effectively 100% of the administered dose by 4–5 days after administration. It should be noted that some material could of course have been absorbed and excreted via the bile into the feces, and this would not have been detected. Baseline measurements of urinary excretion of PVP before giving the 5-fluorouracil, however, showed that 0.013–0.04% (average 0.03%) of the dose was actually absorbed and excreted by that route. This increased by 2–20 fold as a result of anticancer drug treatment. The investigation gave no indication of possible traces of PVP retention within the body, but two interesting observations were made. First, they showed that high molecular weight material was preferentially adsorbed to the stools. They did this by incubating PVP (MW about 30,000) with feces for 18 hours and fractionating the supernatant using Sephadex G-200. The peak molecular weight of the supernatant material was around only 10,000, indicating that the higher molecular weight material was still attached to the feces. Second, they showed that the mean molecular weight distribution of ^{14}C-PVP in the urine of patients who had received PVP orally was correspondingly low, with a peak molecular weight on Sephadex G-100 of about 10,000. This implies selective absorption and/or excretion of lower molecular weight material.

DISCUSSION OF STUDIES OF ORAL
ABSORPTION OF PVP

The evidence clearly suggests from the experiments so far described that a component of the PVP preparations which have been studied is absorbed from the gastrointestinal tract into the body. There is ample evidence from the appearance of radio-labelled material in bile, urine, tissues, and even to a small extent in the expired air, that this can occur. What is at issue is the nature of this component and the rate and extent of absorption. Any discussion of the evidence provided by the individual experiments must therefore start from a consideration of the physicochemical nature of commercial PVP itself. It is not a precisely defined substance. It is composed of molecules of different sizes ranging from monomer through to molecules perhaps 2–3 times the size of the average molecular weight under consideration. There are also small batch variations in the scatter of molecular sizes (see Chapter 1). The difficulty in interpretation of most of the experimental work which has been performed is that it does not enable the nature or size of the absorbed material to be identified. On purely theoretical grounds it is highly unlikely that substantial gastrointestinal absorption of PVP can occur. Indeed, if one considers the evidence relating to pore size, no molecules greater than molecular weight 2,000 can enter the body by this route. Early work by Hober and Hober (1937), for example, suggested that non-lipid-soluble substances with a radius greater than 0.4 nm (corresponding to a molecular weight of 180) cannot pass through the small intestine of the rat. This finding was confirmed by Lindemann and Soloman (1962) who investigated pore size in the luminal surface of jejunal cells in the rat. Fordtran et al. (1965), investigating pore size in the gut in humans, came to a similar conclusion. They showed that effective water pore size in the human jejunum was between 0.67 and 0.88 nm, and in the ileum between 0.30 and 0.38 nm. Schwartz (1981) calculated on the basis of Stokes radii for PVP, but without taking into account convection, intrapore friction or steric hindrance, that molecules up to 2,000 molecular weight might be absorbed.

This would seem to exclude pores as a means by which PVP molecules, other than very small oligomers or monomer, can enter the body. However, the experiments of Loehry et al. (1970), measuring absorption of different molecular weight material from the perfused rabbit intestine, and of Siber et al. (1980), in which the ratio of different molecular sizes of PVP in the administered dose and in the collected urine were analyzed, both showed absorption of higher molecular weight material in the range of 10,000 to 50,000. How then does this higher molecular weight material enter the body?

There are at least two possible explanations:

1. The experiments for estimation of pore size described earlier were not sufficiently sensitive to discover the presence of a few pores of larger diameter.
2. A different mechanism exists for the transport of large molecular weight material through the intestinal wall. The experiments of Clarke and Hardy (1969a, 1969b, 1971) in neonatal animals suggest the existence of a pinocytic mechanism, the importance of which dwindles as the animal matures. Perhaps a few cells retain this ability in the adult. Having said that, it should be noted that these authors could not find labelled PVP in the blood even in neonatal animals where the pinocytic mechanism was operating at its most effective. This may indicate that an inadequate period had been left before blood sampling, or it may point to a difference between the rat and other species, especially since one of these authors had previously shown under similar circumstances that PVP does enter the circulation in the neonatal pig and calf (Hardy 1965, 1968).

SUMMARY OF MAIN FINDINGS

1. The majority of orally administered PVP can be recovered from the feces. The actual amounts range from 60–71% in the ewe to approaching 100% in man and most other species.
2. The majority of the PVP excreted in the feces in both man and the rat is recovered within the first 24 hours.
3. Large doses of PVP given orally cause diarrhea and so may further reduce the availability for absorption.
4. There is preferential adsorption of high molecular weight PVP to feces in man.
5. PVP is absorbed into cells of the distal intestine of weanling rats by pinocytosis, but there is no evidence of transfer into the blood. This is probably part of the physiological process by which neonates absorb high molecular weight immunoglobulins from colostrum. In adult rats, less than 1% is retained in this way.
6. In man, the small amount of PVP which is absorbed following oral administration appears to be absorbed mainly in the upper bowel.
7. Considerations of pore size would suggest that only PVP with a molecular weight less than 2000 would be freely absorbed. Experiments involving perfusion of the rabbit intestine suggest that material with a molecular weight up to 50,000 can be absorbed to a limited extent. There is an inverse log/log relationship between molecular size and absorption.
8. The extent of oral absorption, as shown by measurements of urine, tissue, and biliary excretion, is related to molecular size. With PVP K-30 the total which can be accounted for in this way in both man and the rat is less than 0.04% of the dose.
9. In toxicity studies in which large doses of PVP were fed in the diet for up to 2 years, there were no significant changes attributable to PVP in the rat or cat. A two-year feeding study in the dog revealed slight swelling of the

reticuloendothelial cells in the mesenteric lymph nodes, which may have been evidence of PVP absorption, but this was not confirmed.

A composite diagram summarizing the evidence on absorption, distribution and excretion of orally administered PVP is shown in Figure 4.

CONCLUSIONS

1. Absorption of PVP into the body in most species studied, including man, is minimal. The absorption that does occur is inversely related to the molecular weight (K-value) of the PVP administered, but even with low molecular weight material it rarely would exceed a few percent.
2. There is no conclusive information about the size of the molecules that can enter the body through the gastrointestinal tract, but it is highly probable that the majority of absorbed material is oligomer or monomer, with only a very small fraction of higher molecular weight material, which may be taken up by pinocytosis.

INTRAMUSCULAR AND SUBCUTANEOUS ADMINISTRATION OF PVP

PVP with an average MW of greater than 25,000 was in the past incorporated into sustained-release pharmaceutical formulations for injection, and was extensively used in Europe to prolong, in particular, the effect of vasopressin in the management of diabetes insipidus. Much of the evidence for the absorption of PVP comes from reports of patients who have received repeated injections of such preparations. The storage of PVP which may follow such repeated administration is considered later in detail, but a review of this litera-

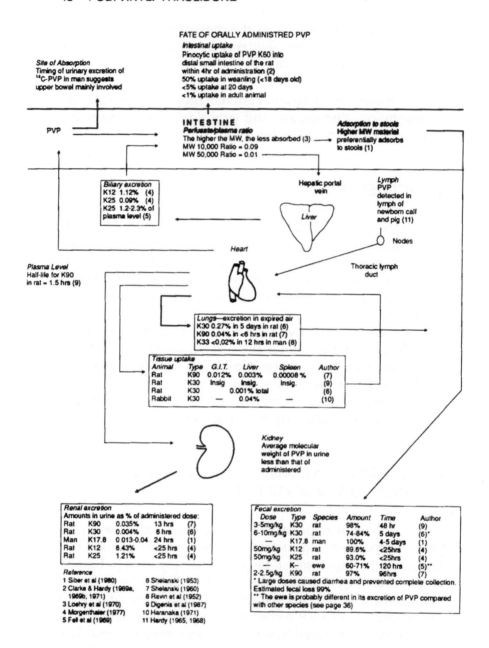

FATE OF ORALLY ADMINISTRED PVP

Intestinal uptake
Pinocytic uptake of PVP K60 into
distal small intestine of the rat
within 4hr of administration (2)
50% uptake in weanling (<18 days old)
<5% uptake at 20 days
<1% uptake in adult animal

Site of Absorption
Timing of urinary excretion of
^{14}C-PVP in man suggests
upper bowel mainly involved

PVP

INTESTINE
Perfusate/plasma ratio
The higher the MW, the less absorbed (3)
MW 10,000 Ratio = 0.09
MW 50,000 Ratio = 0.01

Adsorption to stools
Higher MW material
preferentially adsorbs
to stools (1)

Biliary excretion
K12 1.12%   (4)
K25 0.09%   (4)
K25 1.2-2.3% of
plasma level (5)

Hepatic portal
vein

Liver

Lymph
PVP
detected in
lymph of
newborn calf
and pig (11)

Nodes

Heart

Thoracic lymph
duct

Plasma Level
Half-life for K90
in rat = 1.5 hrs (9)

Lungs—excretion in expired air
K30 0.27% in 5 days in rat (6)
K90 0.04% in <6 hrs in rat (7)
K33 <0.02% in 12 hrs in man (8)

Tissue uptake

Animal	Type	G.I.T.	Liver	Spleen	Author
Rat	K90	0.012%	0.003%	0.00008 %	(7)
Rat	K30	Insig	Insig.	Insig.	(9)
Rat	K30		0.001% total		(6)
Rabbit	K30	—	0.04%	—	(10)

Kidney
Average molecular
weight of PVP in urine
less than that of
administered

Renal excretion
Amounts in urine as % of administered dose:

Rat	K90	0.035%	13 hrs	(7)
Rat	K30	0.004%	6 hrs	(6)
Man	K17.8	0.013-0.04	24 hrs	(1)
Rat	K12	6.43%	<25 hrs	(4)
Rat	K25	1.21%	<25 hrs	(4)

Reference

1 Siber et al (1980)
2 Clarke & Hardy (1969a,
  1969b, 1971)
3 Loehry et al (1970)
4 Morgenthaler (1977)
5 Fell et al (1969)

6 Shelanski (1953)
7 Shelanski (1960)
8 Ravn et al (1952)
9 Digenis et al (1987)
10 Haranaka (1971)
11 Hardy (1965, 1968)

Fecal excretion

Dose	Type	Species	Amount	Time	Author
3-5mg/kg	K30	rat	98%	48 hr	(9)
6-10mg/kg	K30	rat	74-84%	5 days	(6)*
—	K17.8	man	100%	4-5 days	(1)
50mg/kg	K12	rat	89.6%	<25hrs	(4)
50mg/kg	K25	rat	93.0%	<25hrs	(4)
—	K–	ewe	60-71%	120 hrs	(5)**
2-2.5g/kg	K90	rat	97%	96hrs	(7)

* Large doses caused diarrhea and prevented complete collection.
Estimated fecal loss 99%.
** The ewe is probably different in its excretion of PVP compared
with other species (see page 36)

Figure 4.   Fate of orally administered PVP.

ture indicates that PVP is absorbed from the site of administration, and there is evidence for its subsequent appearance in the skin, bone, liver, kidney, spleen and lymph nodes. It should be emphasized that these observations were made after daily administration over many years of large amounts of PVP, e.g., a cumulative dose over a 6-year period of 1200 g (Lachapelle, 1966).

More definitive information has come from a series of experiments in animals performed by Cameron and Dunsire (1983a, 1983b, 1984). Two series of studies were performed in which either 1 mg/kg or 10 mg/kg of ^{14}C-PVP K-12, PVP K-17 and PVP K-30 was injected (in a dose volume of about 12 μl) intramuscularly 4 mm deep into the left flexor digitorum profundus muscle of female Sprague-Dawley rats. In the first series of experiments, eight rats per group were injected and kept in glass metabolism cages for a period of 30 days, collecting urine and feces at not less than daily intervals in order to assess the excretion of ^{14}C-labelled PVP. At the end of this period the animals were killed and the muscle into which the PVP had been injected, and the left popliteal lymph node, were dissected out and weighed, and the ^{14}C-radioactivity was measured. The gastrointestinal tract and carcass were also retained and their radioactivity assessed so that the total recovery of the injected ^{14}C-PVP could be established. The total amount of radioactivity collected from all sources (including cage washings and in some experiments expired carbon dioxide) ranged from 90–97%. Table 7, compiled from the results of these studies, shows that the mean radioactivity remaining at the site of injection after 30 days ranges from 1.7–3.9% of the injected dose for K-30 to 0.12% of the dose for PVP K-12. The majority of the radioactivity was excreted in the urine; the smaller the molecular weight of the PVP, the greater the proportion that appeared in the urine (from 49% for K-30 to around 70% for PVP K-17 and PVP K-12). Conversely, with decreasing molecular size, a greater proportion appeared in the feces (from 10–13% for PVP K-30 to 15–16% for PVP K-12, depending on the dose). In all cases the major proportion (between 84–94%) of the

Table 7.   Percentage Recovery of ^{14}C-PVP 30 Days After Intramuscular Injection

	K-12		K-17		K-30	
	1 mg/kg	10 mg/kg	1 mg/kg	10 mg/kg	1 mg/kg	10 mg/kg
Urine	68.6	70.8	70.7	75.3	49.3	48.6
Feces	16.2	14.6	13.8	11.7	10.1	13.0
Injection site	0.12	0.12	1.4	0.9	3.9	1.7
Draining node	0.00	0.00	0.00	0.00	0.04	0.06
Carcass	0.53	0.15	5.5	6.0	27.0	25.1
G.I.T.	0.02	0.01	0.6	0.4	1.0	0.8
Total recovery	93.2	90.2	94.6	97.2	94.6	91.7

urinary and fecal excretion occurred within the first 24 hours.

In the second series of experiments, the same doses (1 mg/kg and 10 mg/kg) of ^{14}C-labelled PVP K-12, PVP K-17 and PVP K-30 were used, administered intramuscularly as before to a total of 36 animals at each dose level. Four animals were then killed on each of 1, 2, 4, 9, 14, 19, 24, 30 and 78 days after injection, although in the case of PVP K-12 the final measurements were made on day 50 after injection. A 5-ml blood sample was removed from each animal, and the muscle into which the PVP had been injected and the draining lymph node (left popliteal) were removed, taking care to avoid any cross-contamination.

These studies have given important information concerning the rate of disappearance of PVP from the injection site, possible accumulation of PVP in lymph nodes and the levels of PVP detected in plasma. The results are summarized in Table 8. First it was shown that with all three PVP samples the majority of the PVP (86.4–99.7%) was absorbed from the injection site within 24 hours. As might be expected, the removal of the lower molecular weight material, PVP K-12, was greatest, so that only 0.3% of the injected radioactivity could be detected at the injection site after 24 hours. By the end of the study (50 days after injection) this had fallen further, to about 0.1% of the injected dose. In the case of PVP K-17, clearance from the injection site was rather slower, with

Table 8. The Distribution of Radioactivity with Time Following Intramuscular Injection of PVP K-12, K-17, and K-30 in Female Rats[a]

A. Radioactive residues at selected times after injection of 1 mg/kg PVP

Tissue	PVP	Days after injection						
		1	2	4	9	14	30	78[b]
Plasma (ng/ml)	K-12	NS	NS	NS	NS	NS	NS	NS
	K-17	528	243	55	16	15	7	5
	K-30	3185	1921	496	103	91	59	19
Lymph node (ng/g)	K-12	NS	NS	NS	NS	NS	NS	NS
	K-17	878	860	2173	1493	1849	1029	2067
	K-30	2309	3727	3325	7398	3251	10258	4962
Injection site (% dose)	K-12	0.32	0.44	0.35	0.22	0.18	0.12	0.10
	K-17	2.56	2.77	2.27	2.20	2.26	1.38	0.67
	K-30	13.63	11.15	9.38	8.68	7.58	5.72	3.18

B. Radioactive residues at selected times after injection of 10 mg/kg

Tissue	PVP	Days after injection						
		1	2	4	9	14	30	78[b]
Plasma (ng/ml)	K-12	26	10	2	NS	NS	NS	NS
	K-17	4119	2426	741	163	151	92	43
	K-30	33218	22172	9478	1736	912	492	217
Lymph node (ng/g)	K-12	357	839	332	2353	499	1258	514
	K-17	19812	19140	15982	14112	14233	21842	9916
	K-30	28392	41068	61060	62835	61648	77368	37668
Injection site (% dose)	K-12	0.30	0.31	0.28	0.22	0.21	0.13	0.08
	K-17	1.90	1.60	1.71	1.25	1.17	0.74	0.33
	K-30	7.58	6.73	4.49	4.19	4.31	2.41	0.95

[a]Compiled from Cameron & Dunsire (1983a, 1983b, 1984)
[b]In the case of K-12 final measurements were made on Day 50
NS Counts not significantly above background radioactivity

2–3% present after 24 hours. This fell only slightly during the first 14 days of the study to 1–2% of the administered dose. By the end of the study (in this case 78 days) the average level of radioactivity was 0.3% and 0.7% for the 10 mg/kg and 1 mg/kg doses, respectively. For PVP K-30, absorption from the injection site was much slower, with 7.6% of the 10 mg/kg dose and 13.6% of the 1 mg/kg dose present at the injection site after 24 hours. Within 14 days these figures had fallen only to 4.2% and 8.7%, respectively, and even after 30 days, 2.4% and 8.7% remained. After 78 days, 1% of the 10 mg/kg dose and 3.2% of the 1 mg/kg dose were still present at the injection site.

Second, the appearance of the ^{14}C-label in the lymph node reflected closely what was happening at the injection site, so that in the case of PVP K-12, where absorption was rapid, little PVP could be detected in the lymph node even immediately after injection of 10 mg/kg, and within 4 days the levels were undetectable. At 1 mg/kg of PVP K-12, the levels of radioactivity in the lymph nodes were at no stage higher than background values. For PVP K-30 a different picture emerged. Throughout the whole period of investigation the levels of ^{14}C in the draining lymph nodes, following injection of 10 mg/kg, were 10–100 times greater than with PVP K-12 and between 2 and 4 times those seen with PVP K-17. The values remained relatively high throughout the period of study, gradually peaking about 30 days after the injection. Once again, the levels following injection of 10 mg/kg were consistently higher (4–20 times) than those seen following injection of 1 mg/kg.

Finally, these experiments showed that plasma levels of ^{14}C-PVP closely paralleled the retention of PVP within the body. Thus for PVP K-12, where in excess of 90% of the PVP was excreted in the urine and feces within the first 48 hours after injection and less than 0.5% of the PVP could be recovered from the carcass after 30 days, plasma levels were almost undetectable, with only minute amounts measurable at the high dose during the first 2 days. Conversely, with regard to PVP K-30, radioactivity was detectable in plasma throughout the study, especially at the 10 mg/kg dose level. The values were greatest during the first 48 hours, when the maximum urinary and fecal excretion of PVP was occurring; but even after 30 days, significant levels of radioactivity were measured in the plasma. At this time after dosing, the carcass levels were shown to be in the region of 25–27% of the administered dose.

SUMMARY OF MAIN FINDINGS

1. PVP (MW > 25,000) has been used as an adjunct when administering certain drugs (e.g., vasopressin) to delay absorption from the site of injection. These circumstances, repeated daily injections over several years, may cause accumulation of PVP in organs of the reticuloendothelial system, i.e., liver, bone, kidney, spleen, and lymph nodes.
2. In studies in which a single dose of 1 or 10 mg/kg of ^{14}C-PVP (K-12, K-17, K-30) was injected intramuscularly into rats, absorption from the injection site was rapid, with the majority of the PVP excreted in the urine and feces within 48 hours.
3. The rate of absorption is dependent on the molecular weight of the PVP, with up to 4% of PVP K-30 retained at the injection site 30 days after administration. Retention in draining lymph nodes is also observed.
4. Although almost all of the PVP is removed from the injection site, the percentage leaving the body depends on the molecular weight or K-value of the PVP injected. Up to 27% of PVP K-30 is retained within the carcass of the animal 30 days after intramuscular injection. With lower molecular weight PVP, like K-12, the amount retained in the carcass is less than 0.7% of the administered dose, such material being more readily excreted by the kidney.

ABSORPTION OF PVP BY INHALATION

Interest in the possible absorption of PVP by inhalation stems from the description by Bergmann et al. (1958) of clinical and histological pulmonary inflammation, which they attributed to the polymers used in hair spray. They showed X-ray abnormalities in women who used hair sprays daily for 2–3 years which disappeared after discontinuation of use of the hair spray. A lymph node from one of the patients was examined histologically and found to contain unidentified granular phagocytosed material which was assumed to be to PVP. It is this group of workers who gave the name "thesaurosis" to describe the "process of storage" seen in the tissues.

Since then, there have been other similar observations in humans (see the section on storage in Chapter 7), but the role of PVP in these findings is still unclear, mainly because hair sprays contain many other materials such as shellac and vinyl pyrrolidone-vinyl acetate copolymer, any of which could also be stored in lymph nodes. Whether PVP is absorbed significantly by this route into the rest of the body is also unclear from the studies carried out.

In animals, only three studies have been reported, and these are considered in detail in the section on storage (Chapter 7). Rats and guinea pigs were exposed to aerosols of PVP or commercial hair spray, and although minor local changes involving lymphoid tissues were reported, there was no specific evidence for systemic absorption of PVP.

In summary, there is inadequate evidence at present to conclude whether and to what extent PVP is absorbed by inhalation from aerosol sprays.

5

Excretion and Metabolism of PVP

INTRODUCTION

Numerous studies have been conducted in which urinary excretion of PVP of various grades has been measured, the most informative being those where the PVP has been given intravenously. As with the absorption studies, many have been carried out using the PVP simply as a tool to investigate pore size in membranes, and in particular in glomerular membranes. It is evident from these studies that low molecular weight PVP is readily excreted by the kidney, but as molecular size increases the rate and extent of excretion reduce. Above a certain molecular size (about 110,000 MW) no renal clearance of PVP occurs. Commercially available PVP of any particular K-value, although specified to have a particular average molecular weight, does in fact include a range of molecular sizes. If we wish to establish the extent of urinary excretion for molecules of a particular size, it is appropriate to consider data which have been obtained using different PVPs with a range of K-values.

ANIMAL STUDIES (see also Appendixes 2A and 2B)

Wessel et al. (1971) reviewed the world literature on the renal elimination of different PVPs with mean MWs ranging from 11,500 to 500,000 and given in amounts of up to 5 g/kg in a number of different species, including humans. In most

of the early work, PVP excretion into the urine was estimated either chemically or by viscosity measurement, and
although these are recognized as inaccurate and insensitive
techniques they all indicate that PVP is excreted in the urine
at a rate and in amounts that depend on molecular size.
Following intravenous administration, the actual extent of
PVP clearance in the urine during the first 24–96 hours
ranged from 90–95% for PVP with a MW of 12,600 to only
10–20% for PVP with a MW in excess of 100,000. There were
small species differences in the absolute excretion of PVP,
but the overall trend remained the same.

More definitive analysis of renal elimination has come
from studies using radio-labelled PVP, using mostly iodi-
nated material. One of the earliest was carried out by Ravin
et al. (1952), who infused various grades of ^{131}I-labelled PVP
intravenously into normovolemic dogs. By plotting cumulative excretion curves they showed that 90% of PVP K-32,
65% of PVP K-35 and only 15% of PVP K-50 had appeared in
the urine within 72 hours of administration (Figure 5). Since
then, there have been numerous studies with low molecular
weight PVP fractions, all of which indicate extensive and
rapid clearance by the kidney. For example, Hespe et al.
(1977) gave PVP K-18 (98% with a MW of 30,000 or less) and
PVP K-14 (99.5% with a MW of 30,000 or less) intravenously
to rats. They showed that following a dose of 50 mg/kg, 93%
of both PVP K-14 and PVP K-18 was excreted in the urine
within 72 hours (mostly during the first 24 hours). However,
if the dose was increased to 200 mg/kg and observation continued for 22 days, then despite the increased collection
time, the total recovery of PVP K-14 from the urine remained
about the same (92%), while that of PVP K-18 was actually
slightly less (86%).

Using lower molecular weight materials (PVP K-12 and
PVP K-17) Indest and Brode (1977) and Indest (1978) came to
essentially similar conclusions. In their experiments 98% of
PVP K-12 and 94% of PVP K-17 were eliminated during an
observation period of 10–27 days following intravenous
injection into rats (with more than 90% of both materials

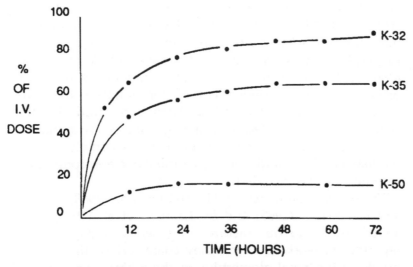

Figure 5. Cumulative urinary excretion of ^{131}I-labelled PVP in the dog following intravenous injection of PVP K-32, PVP K-35, and PVP K-50. (Modified from Ravin, et al. 1952.)

being eliminated in the first 24 hours). Following intramuscular injection, also into rats, of 1–10 mg/kg of three different molecular weight ranges of PVP (K-12, K-17 and K-30), Cameron and Dunsire (1983a, 1983b, 1984) recovered rather less material in the urine over a 30-day period. They collected only 69–71% with PVP K-12, 71–75% with PVP K-17 and much less (about 49%) of PVP K-30 (Table 7). In these studies substantial amounts of PVP appeared in the feces (10–16%), and in the case of K-30, 25–27% was still retained in the carcass after 30 days. However, as with the studies of Indest and his colleagues, the vast majority of the urinary excretion took place within the first 24 hours after injection (84–94%).

Schiller and Taugner (1980), using ^{14}C-PVP K-12 (MW 1,700), have investigated more precisely the actual processes involved in the renal excretion of PVP. Using anesthetized male Wistar rats, they injected intravenously radio-labelled inulin or PVP (150 mg/kg) and measured the levels of radioactivity in urine and plasma at various times after injection. At the end of the experiment the kidneys were removed to measure radioactivity either in homogenates or by autoradiography. First, they found that clearance of PVP K-12 was identical to that of inulin; second, that removal from the bloodstream was independent of concentration over the range 5–10,000 nmol PVP/ml of plasma; and third, that the process was not affected by inducing a diuresis with mannitol, all of which indicates that excretion is entirely by glomerular filtration, and that active tubular reabsorption and secretion can be excluded.

The studies conducted by Hardwicke et al. (1968), Hulme and Hardwicke (1968) and by Lambert et al. (1970) have added further to this picture. In a series of papers they reported on work in which they characterized the sieving profile of the renal glomerulus in the rabbit, dog, rat and humans. By fractionating both the urine and the plasma they were able to assess the glomerular clearance of what was then essentially 20–30 different molecular weight fractions of PVP varying in size by only 0.2–0.3 nm. They plotted the spectrum of permeability for molecules with clearances identical to inulin, i.e., that were cleared at the glomerular filtration rate (GFR), to fractions with a clearance of less than 0.5% of the GFR. It revealed that molecules with radii up to 2.4 nm are cleared as readily as inulin (100% of GFR) but as molecular size increased there was a marked reduction in the ease of excretion until with PVP fractions larger than 6 nm total exclusion occurred. Lambert et al. (1970) concluded that in the dog the glomerular pores range in size from 2.38–6.31 nm, with a mean of 3.86 nm, and that this would allow molecules with a maximum molecular weight of 104,000 to pass into the urine.

This latter observation correlates well with an unpublished study carried out by GAF, reported by Stahl and Frauenfelder (1972) in which 2 female dogs received PVP (MW 38,000) intravenously. Urine was collected over a 36-hour period and was then subjected to fractionation to determine the distribution of PVP of different molecular weights. Small molecules appeared in the urine first, with larger molecules appearing subsequently. The largest molecules retrieved from the urine were in the range of 87,000–104,000.

These figures only apply, however, to clearance of these specific polymers by the glomerulus, for although PVP molecules with a radius of 3.6 nm have a clearance which is 30% of that of inulin, the clearance for albumin, which has a smaller molecular size (3 nm), is only 0.02% of the GFR and even in disease it is seldom more than 1–2% (Hardwicke, 1972). The "effective pore size" is thus dependent on the physical characteristics of the molecule involved, e.g., charge, shape, rigidity, etc.

According to Gartner et al. (1968), studying the renal excretion in the rat of three PVP materials with MWs of 11,500, 110,000 and 650,000, two processes are operating. Molecules with molecular weights up to 25,000 are eliminated rapidly by glomerular filtration through the glomerular capillaries, but larger molecules pass through the postglomerular capillaries into the renal interstitium, and, on the basis of the permeabilities to the PVPs being used, they postulated the existence of three pore sizes. Some of the PVP that passes into the interstitium appears to be removed in the lymph, but the majority passes back into the postglomerular capillaries.

These observations were extended by Vogel et al. (1969), who were investigating the interstitial movement of PVP within the kidney. They measured the lymphatic flow of PVP (MW 110,000) in the rabbit kidney under conditions of normal and furosemide-inhibited tubular reabsorption. They concluded from their data that macromolecules reached the interstitial space of the kidney by diffusion along a concentration gradient and returned to plasma by solvent drag. They also suggested that the interstitial space contains compartments which allow diffusion of macromolecules at different rates. These data, together with previous data from the same laboratory (Gartner et al., 1968), underscore the complexities of renal excretion of macromolecules with molecular size ranges which parallel membrane pore size ranges.

HUMAN STUDIES (see also Appendixes 2C and 2D)

Although intravenously administered PVP was once widely used as a plasma expander, few studies have been conducted in which urinary excretion has been quantitated. This relates mainly to the difficulty of accurately estimating PVP in the urine without the aid of radioactive markers, and the obvious ethical problems of imposing such a hazard on a patient for purely research purposes. Nevertheless, a number of both radioactive and nonradioactive studies have been carried out using amounts of PVP varying from a few milligrams to 10–50 grams. They all confirm that PVP, when given intravenously, is excreted in the urine, and there is some evidence to support the finding in animals that PVP with a lower molecular weight is more quickly and more readily excreted than that with a higher molecular weight.

The first studies were conducted in patients who were receiving PVP as a plasma expander in an attempt to learn more of the pharmacokinetics and possible toxicity of PVP used in this manner. One of the earliest was by Campbell et al. (1954), who used a chemical method for estimation of PVP in the urine and showed that following intravenous infusion of 540 ml of PVP (MW 35,000) 60% was excreted in the urine in 24 hours and 80% within 14 days. They concluded that a multicompartment system was involved in the pharmacokinetics, probably related to the presence of different molecular weight materials in the sample of PVP used.

Wilkinson and Storey (1954), using a similar colorimetric analysis for PVP in urine, investigated the excretion of PVP in 20 patients who had received PVP to maintain the blood volume in various types of shock, injury or following surgical operations. The amounts administered ranged from 500 to 1500 ml of 3.5% PVP solution (Plasmosan) having a nominal MW of 29,000–56,000. They continued analysis of the urine until only trace amounts could be detected (about 260 hours). Within 24 hours, large amounts of PVP had been excreted. In 5 patients excretion was more than 55%, in 13 patients it was between 30–50% and in only two patients was it less than 10%. By the end of the collection period, in 5 patients more than 70% had been voided in the urine, in 9 patients the figure was 50–70%, in 5 patients 40–50%, and it was less than 30% in just one patient.

Excretion of PVP in humans, as has already been demonstrated in animals, is more rapid and more complete with lower molecular weight material. This has been clearly shown in two studies by Brautigam and Gleiss (1956), in which urinary PVP was estimated by chemical analysis. In the first study, 2.7 g of PVP (MW 10,000) was given intravenously to children, and 95% elimination had taken place within 6 hours. In the second study, PVP (MWs 25,000 and 20,000) was given to adults. With the higher molecular weight material, 50% was shown to be eliminated in the urine within 24 hours and 80–85% within 14 days, while with the lower molecular weight PVP elimination was described as "almost complete" within 1–2 weeks. These same authors also showed that the lower molecular weight fractions appeared in the urine first.

A number of studies have been conducted in humans using PVP which has been radio-labelled, but as with other investigations those using iodine labelling are difficult to interpret due to possible unbinding of the label. Loeffler and Scudder (1953) obtained permission from four terminally ill cancer patients to administer 19 g of PVP K-28/32 with an average MW of 40,000 and a probable distribution of molecular weight of 20,000–80,000. The PVP was given intravenously and the urinary excretion followed over 24 hours. At autopsy 2–8 weeks later the tissue levels of radioactivity were also measured. One-third of the radioactivity was excreted in the first 6 hours and two-thirds by 24 hours. The unexcreted portion was distributed widely throughout the body, with the lungs, liver, spleen, kidneys and lymph nodes showing the greatest concentrations.

Longer-lasting studies are more difficult to assess. This was highlighted by Heinrich et al. (1966), who gave 10–20 mg of ^{131}I-PVP (MW 30,000) intravenously and measured the disappearance of radioactivity in the blood and its appearance in the urine. They found that there was radioactivity present in the urine but that not all of it could be attributed to PVP. They suspected that free ^{131}I was actually liberated following uptake of PVP by the reticuloendothelial system (see the section on distribution in Chapter 6).

Siber et al. (1980), as part of a wider investigation on the effects of a cytotoxic drug, used PVP to measure the permeability in the gastrointestinal tract during treatment of metastatic colon carcinoma. Their study showed peak urinary excretion of PVP during the first day after the oral administration of ^{14}C-PVP (MW 11,000), but because the amounts actually absorbed were so small (they recovered approaching 100% of the dose from the feces), the amounts excreted were minute (0.01% of the administered dose). The molecular weight distribution of the PVP in the urine was found to be lower than that in the original sample, an observation which had also been made by Ravin et al. (1952) some years earlier in samples collected during the first 6 hours after PVP administration. This would suggest, as was shown in animals, that the smaller molecular weight components of the PVP material are more easily absorbed and excreted.

More definitive attempts to quantify the size of PVP molecules that could be excreted by the human kidney were made by measuring pore size in the renal glomerulus by Hulme and Hardwicke (1968) and Hulme (1975) in a series of experiments similar to those conducted in animals. Radioiodinated PVP was injected intravenously and serial 1-hour plasma and urine samples were collected. The samples were fractionated to separate the various molecular weight materials and to enable individual clearances of the different-sized PVP molecules to be calculated. These studies revealed that healthy people can excrete PVP molecules with a molecular radius as high as 6 nm, which corresponds to a molecular weight of around 94,000. Slightly lower figures have been indicated by Blainey (1968), who suggested that the glomerulus is highly permeable to molecules with a molecular weight of less than 30,000 but is relatively impermeable to molecules with a size above 70,000. This is in agreement with the work of Robson et al. (1973), using essentially the same technique, who calculated that the maximal pore size was about 4 nm.

On the other hand, the work of Lambert et al. (1970) has shown a slightly higher figure, with glomerular pore size ranging from 1.78–6.69 nm and with a mean of about 3.44 nm. This would allow passage of PVP molecules with a molecular weight of up to 8000 at 100% of the GFR, while molecules up to a maximum of 116,000 could be cleared, but at a much slower rate. In renal disease, increased permeability to PVP may occur, but the extent depends on the nature of the disease (Blainey, 1968). In nephrosis, for example, much larger molecules can appear in the urine (Robson et al., 1973).

DISCUSSION

PVP with a low molecular weight, for example PVP K-12, which has an average MW of 1,700, is readily and almost completely excreted through the kidney, and it would appear that molecules with an effective radius of less than 2.4 nm will be cleared at the glomerular filtration rate. On the other hand, if the injected PVP contains a sizable proportion of molecules with a molecular weight above 110,000, then only the smaller molecular weight material will be cleared by the kidney. None of the molecules above about 110,000 will be removed by this route. At the intermediate sizes, clearance is reduced, so that with a radius of 3.6 nm it is about 30% of the GFR. It has been suggested that such molecules may not pass through the glomerulus but rather through the postglomerular capillaries into the renal interstitium. There is evidence for removal of at least some of this material back into the capillaries or into the lymph.

What then is the significance of these observations to the fate of orally or parenterally administered PVP? First, with regard to oral PVP, it must be remembered that uptake from the gastrointestinal tract is only minimal, and most evidence suggests that the majority of the material entering the blood-stream will have a relatively low molecular weight. The only estimations of the molecular size of PVP to be found in plasma after oral administration have been those of Loehry et al. (1970). The largest molecular weight they recorded in plasma was 50,000, but then only in minute amounts (<0.01% of that in the intestine). Even PVP with a MW of 10,000 entered only to the extent of 0.09% of the amount used to perfuse the intestine. One may conclude that, fol-lowing oral administration of K-30, the material being pre-sented to the kidney will have mainly a low molecular weight and will be present in trivial amounts. Most evidence suggests that this would be readily excreted.

With regard to parenterally administered PVP, urinary clearance is directly related to molecular size of the PVP used, and because the large molecular weight material is not excreted by the kidney, significant retention in the tissues may occur. This was the case when large molecular weight PVP was used as a plasma expander. It is for this reason that the PVP recommended for pharmaceutical use by injection has an upper limit of K-18, which has an average MW of only 9,200. (See discussion [p. 102] in Chapter 7.)

SUMMARY OF MAIN FINDINGS

1. The urinary clearance of PVP has been investigated fol-lowing intravenous administration to animals and man.
2. The rate and extent of clearance of PVP through the kid-ney are dependent on molecular size. The range of molec-ular weights of PVP appearing in the urine is often lower than that of the administered material, showing selective excretion.
3. Large molecular weight material is excreted slowly and to a minimal extent, whereas small molecular weight mate-

rial is excreted more rapidly and more completely (e.g., for K-12 in the rat, 98% is recovered in the urine).

4. The largest molecules of PVP detected in the urine were in the range of 87,000–104,000.

5. Estimates of the maximal pore size in the glomerulus in the dog (2.4–6.3 nm) and in humans (1.8–6.7 nm) are in overall agreement with the size of PVP molecules that actually appear in the urine.

6. For low molecular weight material (MW< 25,000) excretion is probably entirely by glomerular filtration. Some higher molecular weight material may pass through the postglomerular capillaries into the renal interstitium and be reabsorbed.

METABOLISM OF PVP

Three studies have been conducted. They all point to minimal metabolism of PVP, but suggest that it is only low molecular weight fractions, possibly even PVP monomer, which are affected.

The first study was by Ravin et al. (1952), who gave ^{14}C-labelled PVP K-33 by intravenous infusion to humans (three experiments). They found that 0.15–0.2% of the administered dose was expired as ^{14}C carbon dioxide in the first 12 hours, and in the subsequent 12 hours this fell to 0.01%. After 36 hours no radioactivity could be detected. However, dialysis of the original material against running tap water for 48 hours before administration prevented the loss of radioactivity in the expired air, suggesting that a low molecular weight material was responsible.

Shelanski (1960), in a study on a single rat, gave 5 g of ^{14}C-labelled PVP K-90 orally. He showed that 0.04% of the administered radioactivity could be detected in the expired air during the first 6 hours, but after that during a total of 120 hours of observation no further radioactivity could be detected.

On the assumption that at least some of the low molecular weight material in a sample of PVP will be monomer (commercial PVP K-30 contains < 0.2% monomer) a study by McClanahan et al. (1984) is also relevant. They administered ^{14}C-labelled N-vinyl-pyrrolidone (NVP) intravenously into the jugular vein of four rats. The animals were kept in glass metabolism cages and urine, feces and expired carbon dioxide were collected. The total radioactivity in the expired air as ^{14}C-carbon dioxide collected over a 45-hour period was 3.5% of the administered dose, with 1.2% appearing in the first 6 hours and a further 1.2% in the next 12 hours. When urine and bile samples were analyzed following administration of the monomer, less than 0.2% of the radioactivity in the urine and only 2–3% of the radioactivity in the bile was due to unchanged monomer. This indicated that extensive metabolism had occurred, although they were unable at that time to identify the metabolites.

BILIARY EXCRETION

While it is clear that most orally administered PVP is excreted in the feces and most intravenously administered PVP is excreted in the urine, there is evidence that PVP can be excreted in small amounts by other routes. It is important to establish the extent of PVP excretion in the bile, because when considering excretion data from experiments in which PVP has been given orally, some of the activity found in the feces may have come from this source. Data from four investigations are relevant.

One of the earliest reports was by Ravin et al. (1952), who administered ^{14}C-labelled PVP K-33 intravenously to dogs and to humans. They showed that approximately 0.5% of the administered dose appeared in the stools within 24 hours, and that thereafter the fecal excretion fell to about 0.01%/day. In the dog it was possible to confirm by cannulation of the bile duct that all of the PVP lost in the feces was in fact of biliary origin.

More recently, similar conclusions have been drawn regarding biliary excretion in the rat. Morgenthaler (1977), using anesthetized rats in which the bile duct had been cannulated, administered ^{14}C-PVP (K-12 and K-25) intraduodenally. He showed that 1.12% of the dose of K-12 and 0.09% of the dose of K-25 were excreted in the bile over a 20- to 25-hour period.

The situation in the sheep, on the other hand, seems to be rather different. Fell et al. (1969) gave ^{131}I-PVP intravenously to two ewes, and 48 hours later when the small molecular weight material had been cleared by the kidney (i.e., plasma levels had stabilized on the exponential part of the decay curve), they collected the bile and pancreatic fluid flowing from the combined bile duct. This showed that between 1.7% and 2.3% of the plasma ^{131}I-PVP was cleared in this manner. However, they showed by fecal collection, both during this period and at a later stage after diverting the bile back into the gastrointestinal tract, that this contributed insignificantly to the total ^{131}I-PVP lost in the feces (about 4.3% of the intravenous dose), which they concluded passed out through the intestinal wall. This is clearly very different from the situation seen in the dog and the rat, and one must conclude that there are wide species differences.

On the assumption that some material in commercial PVP is monomer, McClanahan et al. (1984) administered vinylpyrrolidone intravenously to rats and showed that 17–20% of the original radioactivity could be recovered from the bile, but that only 2–3% of this was intact N-vinyl pyrrolidone. They also showed that the amount of radioactivity in the feces was considerably less (0.4%) than that excreted in the bile, indicating that enterohepatic circulation of the NVP and/or metabolites was taking place.

SUMMARY OF MAIN FINDINGS

1. Limited studies in humans and rats in which ^{14}C-labelled PVP was administered intravenously show that metabolism is minimal (< 0.3%) and involves only low molecular weight fractions and residual monomer. Some radioactivity was detected in expired air.
2. Studies conducted in the rat, dog, ewe and humans, in which ^{14}C-labelled PVP was administered intravenously, show that small amounts of PVP are excreted in the bile. In dogs and in humans, less than 0.5% of the administered dose was detected, in the rat 1.12% of PVP K-12 and 0.09% of K-25 were found, while in the ewe slightly larger amounts were excreted in the bile (1.7–2.3%).

6

Distribution and Storage of PVP

INTRODUCTION

Occasional reports in the medical literature of excessive storage of PVP in the tissues of patients have raised interest in the question of how and under what circumstances such storage can occur, and how it can be avoided. Because of the particular interest in this topic, a separate chapter has been devoted to analysis of the studies carried out that relate specifically to the storage phenomenon with PVP. The fact that PVP with a high MW, when given in large amounts intravenously, can be stored in the body, is well established. The observation stems from the German medical literature in the 1940s and 1950s, which showed the presence of foam cells or globular deposits in the liver of humans following the intravenous administration of PVP as a plasma expander. Similar observations were subsequently made in animals following intravenous injection of large amounts of PVP.

"Foam cells" is a term used in histological studies to indicate a cell, normally a macrophage, which has large clear inclusions, giving a foamy appearance to the cell. Such inclusions can be produced by polymers or large molecular weight colloids which are pinocytosed by the cells and coalesce to form the large visible droplets. As will be described in detail below, foam cells can be induced in animals and man by the administration of large amounts of PVP. The number of foam cells produced is proportional to both the total amount and MW of the PVP administered. Because of the pinocytosis involved in the cellular uptake of PVP, described in Chapter 8, it is assumed that the foamy appearance of the cells is due to accumulation of PVP within the cytoplasm. There are, however, no definitive histochemical methods for the identification of PVP so that evidence of PVP storage based on the existence of foam cells or intracellular vacuolation is only indirect and suggestive and not definite proof. Studies using radioactive PVP have clearly shown, however, that storage does occur.

The relevance of the observation of storage of PVP following intravenous administration to its use by other routes, and in particular by the oral route, is open to question for a number of reasons.

First, the PVP was given intravenously and in very large amounts ranging up to 28 liters of 3.5% solution (equivalent to almost 1 kg of PVP). It has been shown in both animals and man that only minute amounts of orally administered PVP K-30 enter the body (and then mainly the small-MW material, which is readily absorbed and excreted), and there would appear to be a plasma threshold below which storage is not observed (at least in animals). Thus the effects of administering such huge amounts of PVP intravenously may not meaningfully reflect what happens with much smaller quantities.

Second, the observations were made in patients in whom a state of shock or injury required the administration of a plasma expander. Such a situation might be expected to impair the handling of PVP by the body since, among other things, renal perfusion would almost certainly be reduced.

Third, it is possible to obtain evidence regarding storage in humans generally only by postmortem examination of the tissues. In some instances death occurred within weeks of administration due to the effects of the condition requiring treatment, but in other studies the time lapse was three years or more. In these situations positive identification of PVP in the tissues was not possible, and the researchers relied on histological changes such as globular deposits, vacuolation or the presence of foam cells to indicate that there was storage of PVP.

Despite these reservations, it should be emphasized that in none of these investigations was there a report of any functional disturbance either directly in terms of physiological changes or indirectly as shown by various clinical tests. Furthermore, PVP has been administered therapeutically as a plasma expander to at least 500,000 people without recorded instances of untoward effects attributable to PVP or to PVP storage.

Notwithstanding these observations, it is important when considering data on PVP distribution and storage to establish:

1. What blood levels can be achieved after oral or parenteral administration.
2. Whether storage occurs only if a certain plasma threshold level is exceeded.
3. To what extent the rate of clearance by the kidney affects this process.
4. Into which organs PVP is distributed.
5. To what extent storage is dependent on the molecular size of the PVP.
6. Whether PVP taken up by cells in the body is subsequently liberated or whether long-term storage occurs.
7. Whether storage is associated with functional changes in the cells or organs concerned.

Very few of the available studies have addressed these particular questions, but answers to at least some of them can be gleaned from the existing literature. The investigations that have been carried out fall into four quite distinct types.

1. pharmacokinetic studies
2. distribution studies
3. histopathological and toxicity studies
4. human studies

PHARMACOKINETIC STUDIES OF PVP (see also Appendix 3A)

The elimination of PVP of varying molecular weights following parenteral administration has been tabulated from the world literature up to the 1960s by Wessel et al. (1971). They reported that the half-life ($t_{1/2}$) for the elimination of PVP with an average molecular weight of 40,000 ranges from as low as 12 hours to as high as 72 hours in experimental animals.

One of the most extensive investigations into the plasma half-life of PVP was carried out by Ravin et al. (1952) using ^{14}C-labelled PVP. Following intravenous injection into normovolemic dogs of a number of different-MW materials, they showed that blood levels fell rapidly during the first hour or two. This was followed by a short inflection phase and then a much more prolonged straight-line decay. The rate of disappearance was related to MW, so that at 12 hours after administration 75% of PVP K-50, 40% of PVP K-37, 25% of PVP K-32 and 15% of PVP K-28 still remained in the circulation (Figure 6).

Tokiwa (1958) reported that the $t_{1/2}$ for PVP (average MW 12,600) administered intravenously to rabbits was 4 to 6 hours and that excretion was less in rabbits when the kidneys were damaged by mercuric chloride. Owen et al. (1975) reported that ^{125}I-PVP (average MW 3000–4000) had half-lives of 4.3 ± 1.5 hours and 176 ± 39 hours following intravenous administration to sheep.

Digenis et al. (1987) administered ^{14}C-labelled PVP K-30 by gavage to four groups of five rats. Animals were killed at 6, 12, 24 and 48 hours and measurements of blood and tissue radioactivity at these times revealed only background levels which did not differ significantly from untreated controls. A minor amount of radioactivity (0.04% of the administered dose) was detected in the urine of the rats killed at 6 hours. Nearly all the remaining activity (98.4%) was found in the feces by 48 hours. In a single rat given ^{14}C-PVP K-30 by

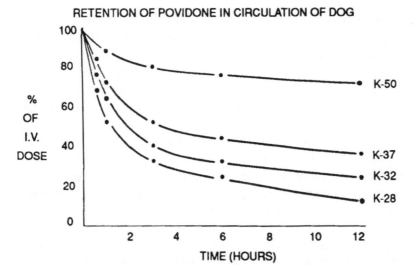

Figure 6.   Relationship between K-value of ^{14}C-labelled PVP and the rate of elimination of PVP from the plasma following intravenous injection into normovolemic dogs. (Modified from Ravin et al., 1952.)

gavage, hourly blood samples were taken for 6 hours and showed that a small amount of radioactivity could be detected at two hours and approached background activity by 6 hours. The authors presented some evidence which suggested that the absorbed radioactivity was probably a low-MW oligomer (less than 3,500) which was present in the original sample.

In the earlier studies referred to above, Ravin and his colleagues (Ravin et al., 1952) concluded from the persistence of PVP in the blood after the initial rapid fall that more was involved in its removal than simple urinary excretion. One possibility, they concluded, was removal into lymph, and to test this hypothesis they collected lymph from thoracic lymph duct fistulae in dogs after intravenous injection of PVP. The PVP appeared in the lymph almost immediately, reaching peak levels within one hour. This was followed by a slow decay phase during which the plasma and lymph levels approached each other. Extrapolation of the decay curves suggested that the lines would intersect (i.e., plasma/lymph equilibrium would be reached) for PVP K-25 at around 9–10 hours and for PVP K-35 after 12 hours. The authors explained the difference in terms of the different MW distributions, suggesting that low-MW material can pass into the lymph more readily.

Additional observations have been made by Vogel and Strocker (1964) who infused PVP (MWs 11,000, 38,000 and 90,000) intravenously into rats and confirmed that passage from the plasma into lymph was dependent on molecular size. The same authors, in another series of experiments (Vogel and Strocker, 1967), infused PVP intravenously (MW 38,000) into rabbits and showed a differential lymph/plasma ratio in different tissues. It was highest in the liver (0.8) and kidney (0.7) and lowest in the hind limbs (0.3) and thoracic duct (0.35). The corresponding ratios for protein were 0.9, 0.8 and 0.7, indicating that, unlike PVP, there was little discrimination between tissues for the transport of protein.

Further support for the concept that PVP passage into lymph is dependent on molecular size came from Boyd et al. (1969), who measured lymph/plasma ratios for lung capillaries using PVP K-11 and K-12 (MWs 3000 and 5000) with molecular sizes of 1.1 nm and 1.7 nm, respectively, administered intravenously to sheep. Also Youlten (1969) studying the rat cremaster muscle, showed, by Sephadex gel-filtration chromatography of the PVP passing out of the blood vessels, that there was differential transport according to molecular size. Table 9 has been adapted from the reported data. The effect of molecular size on apparent volume of distribution is critical between radii of 2.5 nm and 3.1 nm (average MW 16,000–25,000). This is the prevalent range for the PVP preparations most commonly used for experimental studies (e.g., K-30). The apparent volume of distribution ranged from 19.45 ml for PVP with an average molecular weight of 6,500–8,500, to 3.66 ml for PVP with an average molecular weight of 113,500–211,000. This difference in volume of distribution would affect the determination of pharmacokinetic

Table 9.   Effects of Molecular Size of PVP on the Apparent Volume of Distribution in Rats[a]

Estimated Molecular Radius (nm)	Estimated Molecular Weight[b]	Apparent Volume of Distribution (ml)
1.6–1.8	6,500–8,500	19.45
1.8–2.0	8,500–10,500	21.01
2.0–2.5	10,500–16,000	22.99
2.5–3.1	16,000–25,000	13.45
3.1–4.0	25,000–41,500	8.17
4.0–5.1	41,500–67,500	5.43
5.1–6.6	67,500–113,500	4.14
6.6–9.0	113,500–211,000	3.66

[a]Modified from Youlten (1969) – see text
[b]Derived from $s_0 = 0.196M$

parameters. One obvious effect would be to increase the number of apparent pharmacokinetic compartments.

More recently, Vogel (1977) studied the filtration coefficients of PVP with average MWs of 11,500, 25,000, 38,000, 110,000, 160,000 and 650,000 in the rabbit hind limb. He showed that only trace amounts of PVP with MWs of 38,000 and 110,000 were found in the lymph at this site, but that PVP with MWs of 25,000 and greater was found in the liver and kidneys. Furthermore, while PVP with a MW of 650,000 showed significant filtration in the liver (lymph/plasma ratio = 0.14) filtration by the kidney was much less (lymph/ plasma ratio = 0.37).

The lymphatic storage of PVP which has diffused through capillary pores into the lymphatic fluid is attenuated by the fact that the molecule can also diffuse out of the lymphatics. Orally or intravenously administered PVP would be expected to enter the lymphatics in varying amounts, with little following oral administration, due to limitations in intestinal absorption. Following intramuscular injection, a significant fraction of the compound is expected to enter the lymph, because large lipid-insoluble molecules depend primarily on the lymphatic system for removal from intramuscular injection sites (Greenblatt and Koch-Weser, 1976).

Useful information on the pharmacokinetics of PVP has

also come from a study by Regoeczi (1976). He was interested in assessing reticuloendothelial activity, which he did by measuring the rate of removal of ^{131}I-PVP (MWs 33,000 and 36,000) from the circulation following intravenous injection in rabbits. In order to avoid the complication associated with loss of low-MW material through the kidney, he waited 51–56 hours before carrying out any measurements. That way he could be sure he was only measuring the distribution of the large-MW material. At intervals over a 15-day period he measured the plasma and whole-body radioactivities (having first emptied and rinsed out the bladder with saline as an added precaution to avoid possible counting of radioactive material excreted through the kidney). A comparison of the decay curves for these two measurements enabled him to estimate the rate of removal of PVP from the plasma. As a result of various maneuvers to alter phagocytic activity he concluded that PVP was in fact being taken up by cells of the reticuloendothelial system. He also showed that the half-life of PVP in the plasma was 3.2 hours (average) and in the body as a whole 18.0 hours. However, it must be remembered that these figures only relate to the PVP remaining in the circulation after the initial period in which small-MW material was removed by the kidney.

Studies in mice (Carter et al., 1984) using ^{125}I-labelled PVP (MW 30–40,000) have also shown biphasic clearance, with an initial rapid fall for the first 4 hours, followed by a slower exponential fall, corresponding to the phases of distribution and renal clearance and later uptake by phagocytic cells of the reticuloendothelial system. The rate of clearance has also been shown in rats to increase with the age of the animals between 12 and 36 months (Horbach et al., 1986).

It may be concluded that blood levels decline rapidly initially due to removal of small-MW material by the kidney, and that this is followed by a more prolonged decline during which there is passage from plasma to lymph. This process has been shown to be dependent not only on molecular size but also on the nature of the vessel bed being considered.

The ability of PVP molecules to pass out of the bloodstream in this way would clearly affect the volume of distribution.

DISTRIBUTION STUDIES OF PVP (see also Appendix 3B)

A number of studies have been conducted with different grades of radio-labelled PVP given intravenously to different species. Although the timings, doses and procedures have varied, they came to essentially similar conclusions. Three investigations were carried out by Ravin et al. (1952). In the first, rats were injected with ^{14}C-PVP (K-33) at a dose of 350 mg, and pairs of animals were killed at 2, 4 and 7 weeks after dosing. In the second, rabbits were injected intravenously with a total of 8–9 g of the same preparation in seven daily doses and killed at intervals of one, two and six months. The outcome of both studies was very similar. In terms of total amounts of PVP present, the largest fractions in both species were found in skeletal muscle (5–17% of the administered dose) and the skin (5–10% of the dose), which they concluded was due to the large size of these two organ systems. In terms of relative concentrations per gram of tissue, these organs have the lowest amounts, the highest being found in the liver, spleen, bone marrow and lymph nodes. The figures showed that uptake into these reticuloendothelial cells was largely complete within two weeks, but once in the cells the turnover was slow. Most PVP subsequently lost in the urine appeared to come from the extracellular space.

In a further experiment the same authors fractionated ^{14}C-PVP (K-30.2) into high-and low-MW fractions (corresponding to K-37 and K-27, respectively) before intravenous injection into rats (35 mg/kg). The results showed that the overall retention of PVP increased with increasing molecular size, but below a certain molecular size (corresponding to K-26) uptake by the RES was minimal (perhaps due to rapid urinary excretion).

In a somewhat shorter experiment in rabbits, Regoeczi (1976) measured tissue radioactivity 10 days after the administration of ^{131}I-PVP (MW 36,000, but with the low-MW material removed). Expressing the radioactivity as a percentage of that present in the carcass as a whole, he showed that 17.5% was still present in the blood, 21% in the skin, 15% in the liver, and 10% in the bone marrow, with lesser amounts in descending order in muscle, kidney, fat, intestine and spleen.

A similar picture emerged from a study by Loeffler and Scudder (1953) in the rabbit, following injection of 0.8 g/kg of ^{14}C-PVP (K-28/32). Although there was considerable variation in the extent of organ uptake, the greatest amounts were found in the kidneys, lungs, liver, spleen and lymph nodes.

Using fractionated PVP K-17.8 with an average MW of 11,000, injected intravenously into 3 rabbits, Siber et al. (1980) confirmed the selective uptake by liver and spleen. However, they found that tissue levels of radioactivity had returned to background values within 12 weeks of administration. This would suggest that, at least as far as this lower-MW material is concerned, uptake into the RES is a reversible process.

The extent of uptake by individual tissues appears also to be dependent on the absolute amount administered. Experiments by Marek et al. (1969), who gave rats 100 mg/kg ^{3}H-PVP (MW 30,000) intravenously, showed that by 48 hours after dosing the liver contained 35% of the PVP, the muscles 11%, spleen 6%, kidneys 1.8%, lungs 0.7% and the blood 1.8%. They further showed that with larger doses, although uptake of radioactivity was increased in all tissues, a disproportionately large amount was taken up by the spleen.

Using autoradiographic examination of saggital sections of whole rats, six days after intravenous injection of ^{14}C-labelled PVP K-14 or K-18, Hespe et al. (1977) produced a similar picture of tissue distribution. With PVP K-18, 98% of the material had a MW of 30,000 or less, and 99% had a MW of 40,000 or less. With PVP K-14 on the other hand, 99.5% had a MW of less than 30,000 and 100% had a MW less than

40,000. For both preparations the majority of the radioactivity was excreted within 72 hours. However, when the dose was increased there was noticeable retention of a small fraction of the preparation with higher MW, which they observed was gradually eliminated. They also observed that for the same dose of PVP there was generally lower accumulation of K-14 in all tissues except the kidney and gastrointestinal tract.

Although these experiments in general point to some selectivity of uptake by the reticuloendothelial system, specific attempts to induce uptake into particular lymph nodes has been largely unsuccessful. Studies were carried out (Cameron and Dunsire 1983a, 1983b, 1984) in which three grades of ^{14}C-PVP (K-12, K-17, K-30) were injected intramuscularly (1 and 10 mg/kg) into the leg of rats. Throughout a period up to 78 days, animals were killed and the levels of PVP in plasma, muscle and the popliteal lymph gland (which drains the site of injection) were measured. Three observations can be made. First, most of the radioactivity (about 90%) was excreted within 3–4 days. Second, levels in the muscle fell steadily over the period of observation (78 days) to reach 0.08% of the injected dose with PVP K-12 to 3% of the dose with PVP K-30. Finally, there was no consistent change in the lymph node levels of PVP, which remained relatively constant after the initial uptake. The PVP levels in the lymph nodes were related to the molecular weight: the higher the MW the more that could be detected, once again reflecting relative excretion rates. The levels of K-12 were in fact so low throughout as to prevent reliable measurement. These results are discussed in greater detail in the section on absorption.

DISTRIBUTION OF PVP AS MEASURED IN HISTOPATHOLOGICAL AND TOXICITY STUDIES
(see also Appendix 3C)

In 1949 Bull et al., referring to the work of Hecht and Weese (1943), stated that, since PVP used for intravenous infusion will contain polymers large enough to be retained by the glomeruli, and since the polymers are unlikely to be biotransformed, some storage should be expected.

Evidence from toxicity studies to show PVP accumulation in the tissues following oral administration is contradictory (see Chapter 9). It is only from experiments in which the PVP was given intravenously that specific tissue changes have been noted, and then generally this has followed injection of large amounts, often given over a long period of time. For example, Ammon and Muller (1949) injected PVP intravenously into rabbits over several weeks or months until the animals had received a total dose of about 30 g. Histological changes "characteristic of lipid storage disease" were noted, which in animals observed for up to 41 weeks after the end of treatment showed no signs of reversal.

Nelson and Lusky (1951) obtained similar findings. They administered 16 intravenous (ear vein) injections of 10 ml/kg of 3.5% PVP (Periston) in distilled water to 4 male and 2 female rabbits over a 2-month period. There were no systemic effects, but slight splenic enlargement was observed at necropsy and foam cell storage phenomena were observed in the spleen, lymph nodes, bone marrow, adrenal medulla, liver, lungs and thymus (i.e., distribution involved the reticuloendothelial system). Also observed were "slight testicular atrophy, slight epithelial swelling and rarefaction of the distal portions of the proximal convoluted renal tubules and in the choroid plexus of the brain . . ."

A similar approach was used by Frommer (1956) in mice, using three grades of PVP of average MWs 20,000, 40,000 and 125,000. They injected intravenously 0.5 ml of a solution containing between 3.5% and 20% PVP on a number of occasions over a six-day period and examined the liver for histo-logical changes at intervals up to 100 days after completing the course of injections. They found that "foam cells" develop in the liver only following multiple dosing, and then only after a certain critical threshold quantity of PVP has been administered (corresponding to a total body content of 0.1 g/kg). With PVP MW 40,000 this did not occur until the volume of PVP solution administered exceeded the blood volume. Comparative histology showed that the number of foam cells produced by equivalent amounts of low, medium, and high molecular weight PVP was in the ratio 1:2:6.

Ammon and Depner (1957) extended these observations in mice, rats and rabbits, using PVP with average MWs of 12,600, 30,000, 37,000 and 500,000. They reported that tissue retention of PVP was dependent on molecular size, with retention occurring of the higher molecular weight material.

Upham et al. (1956) administering correspondingly large amounts (75 ml of 3.5% solution by infusion) of PVP (MW 40,000) to rats, demonstrated the same type of storage on histological examination. A foamy appearance was noted in the endothelial cells, tubular epithelia and glomerular epi-thelia of the kidney, in Kupffer cells in the liver and in mac-rophages in the adrenal cortex.

In a longer-term study in rats, two PVPs with mean MWs of 34,000 and 270,000 were administered intravenously over a three-day period at doses in the range 0.5–2.5 g/kg (Mohn, 1960). Fluorescence microscopy revealed intracellular mate-rial (assumed to be PVP) in the liver, spleen, lung and kid-ney throughout the one-year period of observation.

Interesting studies on the mechanism of PVP-induced granulomas in rat liver have been reported by Van den Bogert et al. (1986). They used an experimental system that involved continuous intravenous infusion of PVP (MW 40,000). This normally caused the formation of numerous large granulomas composed of clusters of phagocytic cells which contained vacuoles that were thought, on the basis of histological appearance, to be filled with PVP. Under the conditions of their experiments, normal Wistar rats usually died within 2 weeks of continuous treatment (PVP doses of up to 200 mg/kg/day continuously). On the other hand, athymic (nude) rats were more resistant and survived for at least 5 weeks. At that stage the livers of the athymic rats were heavily infiltrated with phagocytic cells, but rarely showed granulomas. Transfusion of thymocytes into the athymic rats, however, led to extensive granuloma formation. This, together with other observations involving the use of doxycycline, led the authors to conclude that the formation of granulomas induced by PVP is a process which is mediated by T-lymphocytes. The thymus-independent actions of PVP are discussed later, in Chapter 8.

A comparative carcinogenicity study in the rabbit by Oettel and von Schilling (1967) has provided still further evidence of PVP storage. Solutions of PVP (5–20%) K-20 (MW 12,000), PVP K-30 (MW 28,000) or PVP K-60 (MW 220,000) were given intravenously at monthly intervals up to one year. Some of the animals were killed after only a few hours, while others were allowed to survive for several years or until the natural death of the animal (up to 7 years). While only isolated accumulations of PVP were found histologically in the liver and spleen with the lowest-weight material (K-20), accumulations were noted not only in these organs but also in the heart, lungs and adrenals with the higher molecular weight PVP samples (K-30 and K-60).

It would seem therefore that histological changes can only be induced if large amounts of PVP are administered intravenously for long periods, or if very-high molecular weight PVP is injected intravenously. Certainly in one well controlled study (Merkle et al., 1983) in which PVP was injected intramuscularly with the specific intention of investigating the histology of the muscle and of the lymph nodes draining the site of injection, there was a notable lack of histological change. Various grades of PVP (K-12, K-17, K-30 and K-90) were injected into the flexor digitorum longus muscle of Sprague-Dawley female rats at a dose of 0.2–2.0 mg on 1 or 5 occasions. The animals were killed 3, 14 or 45 days after the last injection. Local muscle damage following the injection of all materials including saline (control) was noted on the third day and a granulomatous reaction developed in the muscles of some animals (again including the controls) at a later stage. This was not thought to be PVP related. More interestingly, however, the regional lymph nodes showed no specific reaction that could be attributed to the administration of PVP, and there was no evidence of PVP accumulation. These findings correlate well with the distribution studies described earlier in which ^{14}C-PVP was administered intramuscularly and which showed that, although PVP K-30 could be found in the muscle (1–4% of the injected radioactivity) 30 days after injection, only trace amounts (0.04–0.06%) of radioactivity could be detected in the draining lymph nodes (Table 8).

SUMMARY OF MAIN FINDINGS

1. Storage of PVP in tissues of rats, dogs and humans has been reported following the intravenous administration of very large amounts of PVP. Presence in tissues could only be inferred due to lack of specific analytical procedures for PVP.
2. The rate of disappearance of ^{14}C-PVP administered intravenously to dogs is inversely related to the molecular weight.
3. The elimination half-life of PVP (MW 40,000) administered parenterally to experimental animals varies from 1.5 to 72 hours.
4. The removal of PVP from blood appears to be biphasic: a rapid decline due to distribution and renal excretion of low molecular weight material, followed by a prolonged decline due to passage into lymph and tissues.
5. The movement of parenterally administered PVP into lymph and lymph nodes is dependent upon the molecular weight and the dose of PVP administered.
6. Following intravenous administration of radio-labelled

PVP to rats and rabbits, the highest concentrations of PVP are found in liver, spleen, bone marrow and lymph nodes; the highest total amounts are found in skeletal muscles and skin (presumably due to the large masses of these tissues).

7. Histological changes suggestive of storage of PVP ("foam cells") are induced in experimental animals that receive large amounts of PVP administered intravenously for long periods of time. The extent of these changes is dependent upon the molecular weight (greatest storage with highest molecular weight). The aggregation of the storage cells to form liver granulomas may be dependent on T-lymphocyte function.

7

Storage of PVP in Humans

STORAGE OF PVP IN HUMANS FOLLOWING INTRAVENOUS ADMINISTRATION

The early German literature contains reference to PVP storage in humans following therapeutic use of the polymer as a plasma expander (Shallock, 1943; Muller, 1946; Schoen, 1949). For example, in 1946, Muller reported that in infants who had died from "alimentary toxaemia" and who had been treated with PVP intravenously, the spleen and lymph nodes had a pathological appearance which the author associated with storage disease. Loeffler and Scudder (1953) investigated the distribution and storage of PVP following intravenous administration of 6 ml of a 10% solution of ^{14}C-labelled PVP (MW distribution of 20,000–80,000 and an average MW of 40,000) to terminal cancer patients. No histological evidence of PVP accumulation was found on autopsy 2–8 weeks later, although radioactivity was demonstrated in the kidney, lungs, liver, spleen and lymph nodes. Traenckner (1954b) reported on 300 adult patients who had received large amounts of PVP (MW 20,000–80,000) intravenously. Some had received as much as 28 liters of Periston (3.5% PVP) over a 14-day period. Some of the individuals studied were followed for up to 3 years. Persistent storage patterns were observed in the spleen, bone marrow, kidney and liver. In another study of 19 patients (and 51 rats), Traenckner (1954a) reported that similar changes took place in the kidneys after administration of dextran, sucrose and gelatin. Ruffer (1955) examined 20 patients 16 to 39 months after PVP

85

infusions. Neither author reported any clinically detectable changes in kidney or liver function.

There have been studies which show that massive amounts of PVP can cause storage-related functional changes in target organs. Gall et al. (1953) administered a single infusion of 1000 ml of 3.5% to 4.5% PVP solution (MW 40,000; range 20,000–80,000) to 25 patients. This was equivalent to about 0.5 g of the polymer per kg body weight. Serial liver biopsies were obtained prior to infusion and 1.5 to 13 months later. The authors reported that "histological studies of the liver revealed the appearance of basophilic globular deposits ranging up to 50 microns in diameter. These lay within Kuppfer cells or free in the sinusoids and were occasionally accompanied by a mild inflammatory exudate." The deposit rarely appeared before three months and was seen uniformly after six months. The authors questioned the nature of the deposit. A clinical and laboratory followup on 23 of the 25 patients originally treated with PVP was subsequently reported by the same investigators (Altemeier et al., 1954). Observations included: slight or moderate elevation of erythrocytes and hemoglobin, a temporary elevation in the cephalin flocculation test in 11 patients; abnormal retention of sulfobromophthalein in 5 patients; no changes in thymol turbidity, total lipids, or serum bilirubin. There have been other studies of changes in liver function following administration of PVP with generally marginal positive responses (Rheinhold et al., 1952; Rhoads, 1952). The study of Honda et al. (1966) emphasized the large amounts of PVP necessary to affect organ function following intravenous administration. They examined 144 cases in which 3.5% "middle molecular" PVP was used as a plasma expander. They divided the cases into four groups:

1. total use less than 69 g (90 patients)
2. total use of 70 to 99 g (21 patients)
3. total use of 100 to 139 g (20 patients)
4. total use of more than 140 g (13 patients)

Complications were recognized in 49 cases, and included: pulmonary (sputum, hemoptysis, etc.), hepatic (jaundice), renal (proteinuria, hematuria, etc.), intestinal (hematemesis, tarry stool, etc.), wound (wound opening, unhealthy granulation, etc.), and general complications (weakness, anorexia, fatigue, fever, etc.). Dose-response relationships were demonstrated for all complications except the pulmonary ones. The percentages of each group with complications were 17, 43, 65, and 92, respectively. An evaluation of the data presented in the paper indicated that complications due to PVP were limited below a total dose of 69 g but were to be expected with total doses of 70 grams or more.

Kojima et al. (1967) published the most comprehensive morphological study of PVP storage in tissues in humans. They studied 34 autopsy cases from individuals who had received intravenous PVP. In 27 cases, the average molecular weight of PVP given was 24,800, and in 7 cases it was 12,600. Most of the individuals died from neoplastic, cardiovascular or inflammatory disease unrelated to PVP. Storage was characterized by the presence of foam cells and/or massive deposits of vacuolar amorphous clumps. The storage area was generally basophilic in nature. The morphological findings were as follows. With PVP of MW 24,800, with total doses up to 70 g, there was little evidence of PVP storage in any of the tissues. With total doses between 70 and 150 g there was evidence of some PVP storage in liver, spleen, bone marrow and lymph nodes. With total doses above 150 g up to 1015 g there was evidence of marked storage in the liver, spleen, bone marrow, lymph nodes and lungs, and some storage in the heart, tonsils, gastrointestinal tract and urinary bladder.

With PVP with a MW of 12,600 there was only minimal evidence of storage in liver, spleen, bone marrow and lymph nodes, with total doses up to 500 g. These data indicate that storage is dose related as well as molecular weight related and that the relationship of dose to storage is similar to that described by the same group of authors with a different group of patients referred to above (Honda et al., 1966). A number of other observations and conclusions were made. In the kidney, histological changes were primarily manifested as tubular vacuolation in the proximal and distal convoluted tubules and the loop of Henle. PVP storage was also localized at inflammatory loci and surgical wounds. The authors concluded from this that, as a result, ". . . the repair process of the inflammatory foci is markedly retarded, and the defence mechanism against inflammation is also impaired." There are no data which support this conclusion. Similarly, the authors state that ". . . it may be reasonable to assume that heavy PVP storage results in blocking the specific functions of the reticuloendothelial cells and further impairing the resistance of the organism." This assumption is questionable. When larger doses of PVP were administered, it appeared to localize in the areas of neoplastic lesions. They also noted that " . . . widespread dissemination and multiple cancer metastases were recognized in several cases, particularly marked in some cases infused with large doses of PVP." This led the authors to accept this as ". . . evidence that PVP storage promoted the growth and dissemination of already present cancers." That large doses of PVP encouraged metastases is not supported by the data (that PVP might concentrate in neoplastic lesions suggests the possibility that PVP could be useful in focusing the delivery of antineoplastic agents). Tanaka (1971) reported on the autopsies of two patients, one suffering from chronic glomerulonephritis and the other from pulmonary tuberculosis. The first died 39 months after receiving approximately 700 g of PVP (MW 12,600) over a period of approximately 164 days. The second patient received 6–12 g of PVP (MW 12,600) intravenously daily to a total of 721 g. The patient died approximately 2.5 years later. In both patients PVP was identified primarily in the liver, spleen lymph nodes, adrenals and choroid plexus. The authors claimed that ". . . low molecular weight PVP induced renal disorders when administered in large doses." In both patients, the kidneys were ". . . severely atrophic (on account of glomerulonephritis in Case 1 and sclerosis of the arterioles in Case 2) . . ." The rationale behind the authors' conclusion that PVP caused "renal disorders" is unclear. However, it should be pointed out that the interpretation of the paper by Tanaka was based on a translator's version. It is not clear whether disorder implies dysfunction or merely abnormality.

Bubis et al. (1975) emphasized the diagnostic problems which can result from failure to recognize PVP retention in the tissues, and cited a particular case where the storage mimicked congenital mucolipid storage disease. In admonishing pathologists and clinicians on the effects of PVP retention, the authors indicate, for example, that in the case of PVP storage, an ". . . incorrect diagnosis of metastatic mucus-containing tumor may be made by examination of a *frozen section* of an enlarged lymph node" (emphasis added). Leder and Lennert (1972) studied the effects of PVP in lymph nodes and reported that there was an inverse relationship between the level of necrotizing granuloma and the dose administered.

In addition to the above cases, PVP storage, as identified by histological and histochemical techniques using both light and electron microscopy, has been demonstrated in two cases of meningiomas in Taiwanese patients who had previously received PVP as intravenous expanders (Chen et al., 1985).

Fourteen other cases of PVP storage disease, during a 2-year period, have been reported in Taiwan (Kuo & Hsueh, 1984) which were originally wrongly diagnosed as signet-ring cell carcinomas. The authors emphasized the importance of the use of multiple staining techniques to ensure that PVP-containing mucicarminophilic histiocytes should not be mistaken for signet-ring cell carcinomas and lymphomas.

STORAGE OF PVP IN HUMANS FOLLOWING INTRAMUSCULAR OR SUBCUTANEOUS ADMINISTRATION ("DUPONT-LACHAPELLE DISEASE") (See also Appendix 3D)

As mentioned in Chapter 4, PVP with an average molecular weight greater than 25,000 has in the past been used in sustained-release pharmaceutical parenteral formulations. For example, PVP prolongs the duration of action of vasopressin, and was used extensively in Europe in pharmaceutical preparations intended for the management of diabetes insipidus. The use of the preparation involved repeated subcutaneous or intramuscular injections for protracted periods of time. The apparent first report of PVP storage following repeated intramuscular administration was by Towers in 1957. This particular case did not involve the treatment of diabetes insipidus, but dealt with a 28-year-old woman being treated for malignant hypertension with daily intramuscular injections of hexamethonium bromide–PVP. The treatment continued for a period of one year. The patient was adrenalectomized 18 months after cessation of the treatment. At the time of the operation, some enlarged para-aortic lymph nodes were removed. Histological examination revealed changes suggesting the presence of PVP, and the polymer was considered the likely source of the enlargement. There were no reported clinically observed effects attributable to the PVP storage.

Delbarre et al. (1964) reported the first two cases of clinically significant PVP storage following subcutaneous or intramuscular injection in 1964. These involved the administration of vasopressin-PVP preparations for the management of diabetes insipidus in a 55-year-old man and a 42-year-old woman who were treated with daily subcutaneous injections of vasopressin-PVP for 13 and 10 years, respectively. Both developed various signs of bone pain, and radiographic examination showed osteonecrotic changes in the femur as well as other skeletal sites. Biopsies revealed clear amorphous accumulations in bone and in liver of both patients, and in the lungs and ganglia of the male patient. No cutaneous manifestations were reported. The authors concluded on the basis of histochemical staining techniques that the material was most likely to be accumulations of PVP. The apparent first description of the cutaneous effects of protracted administration of PVP was reported by Dupont and Lachapelle (1964), and the same case was described in greater detail by Lachapelle in 1966. It involved a 36-year-old woman with diabetes insipidus who received daily subcutaneous injections of a vasopressin-PVP preparation for about 3 years prior to the development of the first cutaneous lesions. The lesions were described as rounded papules 3–4 mm in diameter and brownish red in color. They were located at the base of the neck in the upper thorax and less densely in the upper limbs and waist. A biopsy of a papule revealed the presence of histiocytes and giant cells loaded with an amorphous substance which was also found in a free state between the cells. It was also occasionally noted in capillary endothelium. Once the lesions had developed, it was another 3 years before the final diagnosis was made, and during that time the patient continued to receive the vasopressin-PVP preparation.

Lachapelle estimated that over a 6-year period the patient had received approximately 1,200 g of PVP (Lachapelle, 1966). Bazex et al. (1966) reported on the case of a 35-year-old woman who had been treated with vasopressin-PVP by intramuscular injection, initially four times a week and then daily, apparently for about 5 years before skin eruptions quite similar to those described above occurred. She also demonstrated some joint symptoms, but these were apparently not associated with PVP storage. Histological examination was similar to that described by Dupont and Lachapelle (1964). Delbarre et al. (1964) referred to the storage as artificial thesaurismosis, and Lachapelle referred to the skin manifestations as cutaneous thesaurismosis. Bazex et al. (1966) suggested that the cutaneous storage syndrome be known as

Dupont-Lachapelle disease ("maladie de Dupont et Lachapelle").[a]

Interstitial PVP has also been identified from skin lesions. Vilde et al. (1968) referred to the accumulation of PVP in the muscle as causing a "pseudo-tumour" in a patient on protracted intramuscular therapy with vasopressin-PVP. Cabanne et al. (1969) described intramuscular injection sites in individuals receiving repeated therapy as showing inflammatory fibrosis which is "hyper-plastic" and "pseudo-tumorous."

The most common clinically significant response to the subcutaneous or intramuscular administration of PVP is the cutaneous storage syndrome. In addition to the reports on the cutaneous effects discussed previously, there have been numerous other reports confirming both the clinical and histological manifestations of this syndrome. Some of these are discussed in detail below, but observations have also been published by Cabanne et al. (1966), Chapuis et al. (1967), Colomb et al. (1970), Duverne et al. (1971), Fartasch et al. (1988), Geniaux et al. (1973), Hizawa et al. (1984), Lachapelle and Bourland (1967), Le Coulant et al. (1967), L'Epee et al. (1969), Rimbaud et al. (1971), Thivolet et al. (1970) and by Uffenorde et al. (1984).

Some reports have emphasized the role of lysosomes in the formation of the lesions (Leung et al., 1970; Bojsen-Moller et al., 1976; Reske-Nielsen et al., 1976). Generally, there are many tissues involved in storage associated with the subcutaneous or intramuscular administration of PVP.

---

[a] "Thesaurismosis" refers to cellular storage of large amounts of endogenous substance due to a metabolic disorder. "Thesaurosis" refers to the excessive storage of endogenous or foreign substance in the body. Thus, it would seem that the latter term is a more accurate description of the applicable situations with PVP. Both terms are outdated. "Storage disease" is used to describe clinically manifested storage of endogenous materials resulting from metabolic disorders. To avoid confusion and misunderstandings of the toxicology of PVP, the following terminology is suggested: Where storage results in a clinically manifested event, e.g., cutaneous papules or bone pain, "storage syndrome" preceded by the name of the tissue (e.g., "cutaneous storage syndrome," "bone storage syndrome") should be used. Where storage occurs but without clinical manifestation, then just "tissue storage," "cutaneous storage," "liver storage," etc., should be used.

These include lymph nodes (Moinade et al., 1977; Widgren, 1965; Caulet et al., 1968; Leder and Lennert, 1972), kidney (Grunfeld et al., 1968; Fonck-Cussac et al., 1970), and liver and spleen (Fossati et al., 1972). Plauchu et al. (1970) described periocular infiltration of PVP in patients following the prolonged use of vasopressin-PVP intramuscularly. Localization in the bone has also been reported and was expressed as a storage syndrome (Vilde et al., 1968). Others have also reported localization of PVP in bone or bone marrow (Faivre et al., 1975; Bojsen-Moller et al., 1976; Reske-Nielsen, 1976; Tereau et al., 1978), some with clinically relevant signs of injury, such as spontaneous fracture (Bojsen-Moller et al., 1976) and joint signs (Faivre et al., 1975). Bert et al. (1972) reported on a case of a 54-year-old woman who had received intramuscular PVP in association with therapy for diabetes insipidus for nine years and a nasal inhalant form of vasopressin-PVP for four subsequent years. It was estimated that she had received approximately 3 kg of PVP. The patient showed signs of an osteoarticular syndrome, though the authors did not verify the presence of PVP in the bone or the synovia.

Although the patient described by Bert and his colleagues (1972) did not show the usual cutaneous manifestations, such as papules, nodules or infiltrate of patches which are usually described following intramuscular or subcutaneous administration of large doses of PVP, the patient did show inflammatory masses in the gluteal muscles at the injection site. The masses required surgical removal, and they were found to be chronic granulomas, with histological appearance typical of PVP localization. This patient also had pancytopenia, and biopsy revealed a complete disappearance of hematopoietic marrow, atrophy of the trabeculae of bone, and occupation of all medullary spaces by PVP-containing histocytes. This appears to be the only hematologic syndrome which has been described in a patient with excessive PVP tissue storage. This patient also displayed hepatomegaly, a condition which was also described in a patient with a similar background by Fossati et al. (1972). The massive amount of PVP stored by the patient described by Bert et al. (1972) also appeared to have an effect on renal function. Creatinine clearance (20.4–29.8 ml/min) was well below normal. This was associated with an infiltration of the renal medullary interstitium by PVP-containing histocytes and tubular atrophy. A number of cases of PVP-induced renal lesions have been reported, but this is the only one with reported diminution of renal function. Most renal changes are manifested as vacuolation of the tubular cells. This commonly occurs with a number of substances which are not degradable by lysosomal enzymes. Doolan et al. (1967) showed that vacuologenic doses of EDTA had no effect on renal function in the rat unless doses were massive enough to cause vacuolation which disrupted the brush border of the tubules. It would appear that this might have occurred in this case, and that this effect, together with the infiltration of histiocytes, resulted in functional renal damage. In a review of published and unpublished clinical information up to 1979, Cabanne et al. (1969) classified the anatomical and clinical characteristics of the storage of large amounts of PVP resulting from the parenteral administration of repeated doses over a protracted period of time. The primary lesions were classified as infiltration of tissues by polymorphic histiocytes, primarily vacuolated, and the interstitial deposit of amorphous or granular substance. Associated lesions included inflammatory reactions which were exudative or granulomatous. This included the observation that ". . . it is not unusual, as in the area of repeated intramuscular injections, that the inflammatory fibrosis becomes hyperplastic and pseudo-tumorous." Site specificity included osteoarticular damage, cutaneous damage, adrenosplenohepatomegalic damage, muscular damage in the injection area, and pulmonary damage. Histiocytosis of the alveolar areas, including the presence of giant cells, has been identified. Pulmonary effects related to inhalation of PVP-containing hair sprays will be discussed at the end of this section.

The summary prepared by Cabanne et al. (1969) accurately reflects the histological and clinical observations made prior to the time of its publication. There have been no new observations revealed in the clinical observations published since then, yet the recognition of tissue storage of PVP, or storage syndrome, is often treated with the alarm of a "new discovery" when it has been observed in recent years. The "pseudotumors" have been a major focus of concern. Though there were initial reports which suggested that PVP is carcinogenic, available data do not confirm this. This is discussed in greater detail in a later chapter. The presence of "pseudotumors" and the confusion concerning carcinogenicity resulting from an incomplete knowledge of the literature can lead to faulty conclusions concerning the health risks of PVP.

The tumorigenicity issue is a semantic problem which is exacerbated by the multilingual nature of science. A tumor may be simply defined as a swelling. Any increase in size of tissue due to space-occupying material or hyperplasia is a tumor. In the current atmosphere of regulation and public concern as reflected by the communications media, and by some scientists, the word "tumor" sometimes refers to a neoplasm. Caulet et al. (1968) referred to the lymphatic swelling observed in one case of an individual having received parenteral PVP as a pseudotumor.[a] Couinaud et al. (1970) reported the case of a 35-year-old man who was treated for 4.5 years, parenterally and nasally, with vasopressin-PVP for the treatment of diabetes insipidus. The patient suffered abdominal pains and other symptoms which led to a possible diagnosis

---

[a]Again, there is a semantic problem which may bear on the interpretation of the statements made. In English, using the prefix "pseudo" in association with a medical or pathological diagnosis carries an implication of "deceptiveness" in the diagnosis. That is, the condition carrying the prefix might be mistaken for the actual state, e.g., "pseudohypoparathyroidism." The use of the word "pseudotumor" carries the same implication. In this case, "tumor" obviously refers to a neoplasm rather than to a swelling. The importance of the distinction can be seen from the paper by Caulet et al. (1968). The paper is in French and contains an English summary in addition to the French summary. The French summary refers to a "pseudotumeur." The authors' translation of that word in the English summary is not "pseudotumor" but "tumor-like." The latter does not have the same implication of misleading diagnosis that the former has. There is a critical diagnostic difference between something that "looks like a tumor" and one that might be "mistaken for a tumor."

of gastric or appendicular perforation. During surgery, a nodular lesion of the greater omentum was discovered which initially was considered to be a tumor. It was found that the nodules were foreign body granulomas which the authors concluded contained PVP. The patient had been previously withdrawn from steroids and the authors considered the most likely cause of the clinical abdominal signs to be adrenal insufficiency secondary to the steroid withdrawal rather than from the PVP-containing nodules in the greater omentum.

Moinade et al. (1977) reported on the case of a woman who was treated with vasopressin-PVP. The amount of PVP administered and the duration of treatment were not specified, but it was apparent that she had received cumulatively large amounts of PVP. She had two painful indurated masses at the injection site, and these masses contained PVP. Three months later she presented with a "large tumor" of both buttocks, though the tumor was not reported to be neoplastic. They observed the presence of PVP storage in the mass. This patient also showed renal dysfunction and had ocular and adrenal lesions.

The amounts of PVP administered by subcutaneous or intramuscular injection cited in the studies above are not always described. It appears that a minimum of 200 g cumulative dose of PVP is required before signs are noticed; with some individuals, as much as 1 to 2 kg cumulative dose is required. An exception to this was found in the report by Gille and Brandau (1975). They reported the case of a 54-year-old female patient who presented with thickening and hardening of the left breast. Ultimately, extensive excision was undertaken and a mass approximately the size of an egg was removed. Granular tissue, foamy macrophages and multinucleated cells were identified histologically. Histochemical studies identified PVP in the mass. The case history showed that the patient had suffered an injury 2 years previously which resulted in pain in her left breast. She injected a preparation containing 100 mg PVP (MW 30,000), 20 mg procaine hydrochloride and 14.2 mg caffeine per ml directly into the breast. The intended use of the preparation was as a prolonged action local anaesthetic. This patient injected a total of 25 ml of the preparation over a period of 2–3 weeks, which would amount to 2.5 g PVP. The resulting foreign body reaction occurred from the retention of the PVP, which would be expected because of the low circulation in the adipose tissue in the breast. The PVP-laden mononuclear cells were those which had reached the site as part of the inflammatory response. It was not an example of the storage phenomenon per se. These findings suggest that caution must be exercised when injecting polymers into an area of minimal circulation.

Hoelscher and Altmannsberger (1982) reported the case of a 69-year-old woman who had received injections of the same preparation on both paravertebral sides in the neck region for 9 years. It was estimated that the patient had received approximately 50 g of PVP. The patient presented with painful swelling in the neck region and the appearance of an apparent tumor which was diagnosed histologically as a malignant xanthogranuloma. This was prior to the knowledge that the patient had received the medication. The tumor showed no nuclear polymorphism nor increase in mitotic rate. Malignancy was suspected because of its size, the large area of necrosis and the deep infiltration into the musculature. Subsequently, PVP was histochemically demonstrated to be present in the lesion. The same issue of *Deutsche Medizinische Wochenschrift* contained an article by Bork (1982) on the so-called pseudotumors associated with PVP. He cited a case of a 48-year-old male who presented with frontal and periorbital tumorous infiltrates. They developed after the patient had received a total of 55 g of PVP from the administration of the PVP-procaine-caffeine preparation in those areas over a 10-year period. The original diagnosis was that of a xanthofibroma with its associated pessimistic prognosis. History, clinical course and further histological examination led to the conclusion that the tumors were a PVP deposit with foreign body reaction. There have been other similar cases of PVP storage and foreign body reactions mistakenly diagnosed as malignant tumors (Soumerai, 1978; Kossard et al., 1980; Wunsch & Kirchner, 1981).

Hoelscher and Altmannsberger (1982) concluded that PVP should not be used as a depot carrier for injectable drugs. Any polymer which cannot be biodegraded at the site and which is given in too large a quantity to allow for removal by the blood supply before the administration of a subsequent amount of material has the same potential for reactivity as PVP. Gaffney and Casley-Smith (1981) repeatedly administered plasma or PVP subcutaneously to rats. The plasma proteins caused chronic inflammation in the skin and fascia to a greater extent than did PVP.

FACTORS TO BE CONSIDERED IN THE USE OF POLYMERS IN INJECTABLE PREPARATIONS

The following should be considered before injecting a depot preparation that contains polymers:

1. the injection site
2. the molecular weight of the polymers
3. the amount of polymer injected
4. the frequency of dosing

Injections into tissues with low blood circulation or muscles with limited mass will delay removal of the polymer. Furthermore, since removal from the site depends upon the circulation, it is compromised if the circulation is not adequate to remove the material injected. High molecular weight compounds such as PVP depend primarily on the lymphatic system for removal from intramuscular injection sites (Greenblatt and Koch-Weser, 1976). Diffusion is inversely proportional to the molecular weight. For example, Sund and Schow (1964) measured the clearance rate of various carbohydrates from rat muscle. Inulin with a molecular weight of 3,000 to 4,000 had approximately 3 to 4 times the clearance of dextran with a molecular weight of 60,000 to 90,000. Mannitol (MW = 182) had a 5-minute clearance, approximately 15% greater than sucrose (MW = 342). Inulin had approximately one-third the clearance of sucrose and approximately 3 times the clearance of dextran. This demonstrated a log dose-response relationship, from which it can be estimated that the clearance rate for a polymer with a molecular weight of 30,000 would be 1.25 to 3 times that for one with a molecular weight of 15,000. The work of Beresford et al. (1957) suggests that this difference in clearance is probably insignificant. They showed that intramuscularly administered iron dextran (MW = 10,000–20,000) was largely absorbed during the 72 hours immediately following injection. During the first 48 hours after administration, there was an increasing diffusion of polymer away from the injection site, with an increasing infiltration of macrophages as part of an inflammatory reaction. After 72 hours, most of the polymer present at the injection site was present within macrophages. One week after the injection there was active regeneration in the injection site.

STUDIES OF THE EFFECTS OF INHALED PVP

Bergmann et al. (1958) described clinical and histological pulmonary inflammation which they attributed to the polymers used in hair spray. This was based on the fact that women who had used hair sprays daily for 2 to 3 years had X-ray abnormalities that disappeared after discontinuation of the use of the hair spray. A lymph node from one of the patients was examined histologically and found to contain unidentifiable granular phagocytosed material. This was attributed to PVP, although the staining characteristics were not clearly those of PVP. Since the first reports of these observations in 1958, there has been substantial controversy concerning the causal relationship between the use of hair sprays and the development of pulmonary toxicity. The issue is clouded by the difficulty of determining whether storage has occurred and what form it takes, and by the fact that hair sprays contain many polymers, including shellac and vinylpyrrolidone-acetate copolymer.

There have been almost 100 publications concerning polymer pulmonary toxicity from hair sprays. Cambridge (1973) reviewed the publications between 1958 and 1972. This included a description of 11 epidemiologic studies of hair dressers, totalling 2,155 individuals. There were 12 cases of tissue storage described—11 of them in a single study (Gowdy and Wagstaff, 1972). Cambridge (1973) suggested that, since there are a large variety of diffuse pulmonary diseases, before labelling a condition as a storage disease there should be some convincing evidence showing identifiable material in the lung or the lymph nodes. For example, sarcoidosis is a granulomatous disease which, on the basis of a Kveim test, may have been mistaken for a pulmonary storage disease (Nevins et al., 1965). Bergmann et al. (1962) discussed other new cases of putative tissue storage syndrome and concluded that there was a strong assumption, but no proof, of an etiological relationship between the inhalation of hair sprays and the storage syndrome. Lowsma et al. (1966a) exposed rats to aerosols of PVP 5 days per week, 8 hours per day, for 30 exposures. Histological sections revealed no inflammatory response. However, macrophages laden with granular material were present throughout the alveoli, and staining results were compatible with the presence of PVP. In another study by Lowsma et al. (1966b), in which the particle size was carefully controlled (0.5–4 μm), similar results were obtained. Cambridge (1973) exposed guinea pigs to aerosols derived from commercial hair sprays for 6 hours per day, 5 days a week, for 1 year. No histopathological changes other than progressive pulmonary lymphoid infiltration were seen. The progressive pulmonary lymphoid infiltration was also seen in the control groups. The data reviewed failed to identify PVP, per se, as a particular threat to health compared to other polymers.

A more recent review of pulmonary thesaurismosis due to

PVP (Ameille et al., 1985) also considered that the true existence of this disease still had to be confirmed.

DISCUSSION CONCERNING THE DISTRIBUTION
AND STORAGE OF PVP IN TISSUES

In the introduction to Chapter 6, a number of questions
were posed regarding the distribution and storage of PVP
within the tissues. From the experiments so far described it
may be deduced that the manner and extent of PVP uptake
into tissues is dependent on a number of factors, such as the
molecular weight of the material used, the site of administration of the PVP, the time of observation after administration
and the amount of PVP given.

The pharmacokinetic studies show that once renal excretion has been taken into account, then removal of PVP from
the plasma is dependent primarily on the transfer into
lymph and on uptake into tissues. In both cases the degree
of transport depends on the tissue involved and, at least in
the case of transfer into lymph, the molecular weight of the
PVP.

There is no evidence to suggest that molecular weight
alters the rate or extent of pinocytotic uptake by tissues. It is
likely nevertheless that in vivo the PVP taken up will be
largely of higher molecular weight simply because the process of uptake is relatively slow and the lower molecular
weight material will be rapidly excreted by the kidney. One
of the most striking demonstrations of this fact was seen
histologically in the experiments of Frommer (1956) using
three grades of PVP (MWs 20,000, 40,000 and 125,000). He
showed that foam cells were produced in the liver in the
ratio 1:2:6 for the low, medium and high molecular weight
material respectively, reflecting the ease with which each
type of material is cleared from the body and thus the
amount of material remaining to be taken up into macrophages. However, it is important to note that he only

achieved such concentration of PVP into a particular tissue (to an extent that became obvious histologically) by repeated injection of large amounts intravenously. Other experiments have illustrated that uptake of PVP is actually a generalized phenomenon which takes place into most tissues in the body, and in absolute terms the highest uptake by single tissues has been shown to occur in the skin and muscle. In terms of relative uptake, however, the specialized phago-cytic tissues still have a greater capacity to take up PVP, and the largest concentrations per gram of tissue have generally been found in tissues which contain a major part of the reti-culoendothelial system, such as liver, spleen, bone marrow, lungs and lymph nodes. On the other hand, when PVP was injected intramuscularly in the rat in an attempt to investi-gate uptake into the lymph nodes draining the site of injec-tion, only a low concentration of radio-labelled marker in the nodes was noted, and although the muscle content fell steadily over the period of observation, there was no corres-ponding increase in the lymph node concentration.

Thus, it is by no means certain that selective concentration of PVP into RES cells occurs unless the renal excretory sys-tem has been overwhelmed by injection of very large amounts of material, and this generally only takes place fol-lowing intravenous injection, especially of higher molecular weight PVP.

It is from a consideration of the above factors that the following recommendations regarding the parenteral use of PVP have emerged.

The German drug regulatory body (Bundesgesundheit-samt, 1983) has issued recommendations concerning the par-enteral use of drugs containing PVP. For intravenous use, drugs may be used containing an unrestricted (but identi-fied) amount of PVP up to K-18. For intramuscular use, each dose of drug should not contain more than 50 mg of PVP, with an upper limit of K-18. Caution is advised on repeated use in patients with poor renal function and the rare possibil-ity of accumulation of PVP in the reticuloendothelial system, and the production of foreign body granulomas should be considered. It is advised not to inject the material into tissues with reduced perfusion, and different injection sites should be used for repeated administration. These recommendations are appropriate, and could be adopted with advantage by others.

The mechanisms involved in the uptake of PVP into reticuloendothelial cells and the consequences of this are discussed in the next chapter.

SUMMARY OF MAIN FINDINGS

1. Storage disease has been reported in some humans who received large doses of PVP (greater than 70 g) administered intravenously. Storage patterns (presence of foam cells and/or deposits of vacuolar amorphous clumps) were observed in spleen, bone marrow, kidney and liver. These same patterns were reported following administration of dextran, sucrose and gelatin. Storage is related to dose and molecular weight of the PVP administered. Marginal alterations to liver function have also been reported following prolonged administration of high doses.
2. Parenterally administered PVP was reported to accumulate in neoplastic tissue.
3. Dupont-Lachapelle disease, cutaneous storage syndrome or cutaneous thesaurismosis are terms used to describe the apparent storage of PVP in humans following subcutaneous or intramuscular administration of large amounts. Storage of PVP has been reported in lymph nodes, kidney, liver, spleen, bone and bone marrow. A minimum cumulative dose of 200 g of PVP is required to elicit signs of PVP storage. Storage of PVP is dependent on the molecular weight and the amount of PVP injected, the frequency of injection and the site of injection (extent of blood circulation).
4. There have been no reported unique pulmonary effects following the inhalation of PVP sprays.

8

Functional Consequences of PVP Uptake by Body Tissues, with Particular Reference to the Reticuloendothelial System (RES) and the Immune System

MECHANISM OF UPTAKE OF PVP INTO BODY TISSUES

The pattern of PVP uptake by various tissues, as indicated by the distribution and toxicity studies, undoubtedly points to removal of material into the major sites of reticuloendothelial cells in the body, e.g., spleen, liver, etc., as discussed in the previous two chapters. It is also clear that extensive uptake occurs in other tissues, e.g., skin and muscle. What is not totally clear is the mechanism by which this occurs.

General considerations

It has already been established in relation to intestinal absorption and renal excretion that only small molecular weight PVP can pass through membrane pores. It is probable, therefore, that uptake into tissues occurs as a result of an endocytic mechanism. Two types of endocytosis have been distinguished, namely phagocytosis ("cell-feeding") and pinocytosis ("cell-drinking"), which is sometimes also referred to as fluid-phase endocytosis. The former may be described as a receptor-mediated vesiculation process in which attachment of particles to the receptor region plasma membrane triggers their incorporation into a vesicle or phagosome. Pinocytosis, on the other hand, is an ongoing process in which very small vesicles (70–100 nm) are pinched off from the cell membrane. It is the mechanism by which a substantial part of the cellular uptake of extracellular fluid takes place. It is thought that in the course of such transport, other soluble material, including polymers such as PVP, may incidentally be engulfed. There is some evidence to suggest that substances adsorbing onto the plasma membrane can act as a trigger – so called receptor-mediated micropinocytosis, and many proteins (e.g., insulin, transferrin, maternal IgG) pass into specific cells in this manner (Houslay and Stanley, 1982).

A third process known as macropinocytosis has also been described which is similar to phagocytosis except that no particles are enclosed.

The outcome of such endocytic activity depends largely on the nature of the material engulfed and on the cell type involved, but a number of events are common. For example, reversed pinocytosis or exocytosis can occur, both on the same side of the cell and on the opposite side (i.e., transport may occur). During passage through the cells two other processes may take place. The pinosomes (or in the case of phagocytosis, phagosomes) may coalesce with each other or they may fuse with lysosomes to form pinolysosomes (or phagolysosomes). If the engulfed material cannot diffuse out and is not digestible by the lysosomal enzymes then accumulation of pinolysosomes (or phagolysosomes) will occur within the cell. The pinolysosomes will then coalesce until a stable (i.e., low surface free energy) membrane state is achieved. This generally results in the accumulation of pinolysosomal vacuoles which remain within the cell for a prolonged period of time. It has been suggested that this may account for the foamy appearance of RES cells exposed to high concentrations of PVP, the so-called "foam cells."

STUDIES ON THE MECHANISMS OF PVP
UPTAKE (see also Appendix 3e)

A number of studies have been carried out which suggest that uptake of PVP (MW 40,000) is by pinocytosis. Roberts et al. (1976), for example, have investigated the uptake of PVP into the visceral yolk sac by injecting it intraperitoneally into pregnant rats at various times after gestation. This tissue was used because it is known to transport material like IgG into the fetus by receptor-mediated pinocytosis. When PVP was injected on the 16th day of gestation and the epithelial cells of the embryonic visceral yolk sac were examined 36 hours later, the cells had a highly vacuolated appearance. A similar effect was produced by Triton-139 (a lysosomal enzyme-resistant detergent) when it was injected at a dose of 0.5 g/kg. The authors suggested that PVP, when used in these large doses, has vacuologenic activity. PVP has also been used in rat isolated yolk sac in vitro cultures to measure the effects of teratogens on pinocytosis (Marlow and Freeman, 1987).

Pinocytic uptake of PVP has also been demonstrated by macrophages. Pratten et al. (1977) incubated rat peritoneal macrophages in vitro with ^{125}I-PVP (10 μg/ml) and showed that it was taken up in a constant linear manner over a 10-hour period. They also showed that if macrophages which had already taken up labelled ^{125}I-PVP over a 3-hour period were then washed and re-incubated in fresh medium, the cells rapidly released between 10 and 50% of the marker into the incubation medium over a 10-minute period. Little further loss of radioactivity took place after this. This implies that much of the labelled PVP was probably loosely bound on the cell surface.

The rate and mechanism of uptake are also dependent on the molecular weight of the PVP (Duncan et al., 1981; Pratten and Lloyd, 1986) so that very high molecular weight material (MW 7×10^6) was captured several times more rapidly than PVP of molecular weight less than 100,000.

Tissue with a high cell content of fixed macrophages might thus be expected to take up PVP in large amounts, and indeed this is confirmed by the tissue uptake studies, which show particularly high levels of PVP in liver and spleen (see earlier). However, there is evidence that such a process can occur at other sites, including the gastrointestinal tract, thyroid, lymph nodes and kidney. The experiments of Clarke and Hardy (1969a, 1969b), for example, describe pinocytic uptake of PVP by intestinal cells in neonatal rats, and although it was evident that such uptake declined markedly as the animals reached maturity, the process was clearly pinocytic.

Schiller and Taugner (1980) described a similar process by cells of the proximal renal tubules, although it should be added that this is by no means peculiar to PVP. Similar uptake has been shown at this site for gelatin, inulin, dextran and various proteins, all without impairment of renal function. By taking autoradiographs of frozen sections of rat kidney 15 minutes after intravenous infusion of ^{14}C-PVP K-12, they showed the presence of radioactivity within the tubules. If the kidney was perfused with isotonic mannitol-NaCl to wash out the tubules, then PVP was apparent within the epithelial cells. If sectioning was delayed until 6 hours after injection, then most activity was found by autoradiography to be in the proximal convoluted tubules. On analysis with high resolution electron microscopy the ^{14}C-PVP was found to be in large lysosome-like bodies of a type frequently observed after administration of macromolecular compounds.

In contrast to this work suggesting uptake of PVP by pinocytosis, Regoeczi (1976) has presented evidence which on the face of it suggests a phagocytic mechanism might be involved. He used a technique (see Pharmacokinetics section) in which he infused ^{131}I-PVP intravenously into rabbits and then after a period of 65 hours (during which elimination of small molecular weight material took place) he compared the radioactivity in the body as a whole with that in the plasma. This enabled him to measure the rate of uptake of PVP into the reticuloendothelial system (or more correctly the rate of removal from plasma). He then used a number of maneuveres designed to alter the rate of uptake of material by phagocytic cells. For example, he showed that intravenous infusion of fibrinogen, a nonimmunological stimulus, caused a 2.6-fold increase in the rate of uptake of ^{131}I-PVP. Creating a sterile subcutaneous turpentine inflammation had a similar stimulant effect, which lasted for a period of at least four days; by that stage the plasma level of radioactivity had declined to such an extent that the level had become undetectable. Regoeczi (1976) also demonstrated that immunological stimuli had a similar effect. He gave an intravenous injection of homologous antiserum to human serum albumin (HSA) to rabbits who had already received ^{131}I-PVP 56 days earlier. He then gave HSA and showed a delayed enhancement of phagocytic activity some 12 hours later, i.e., during the period of maximal acute phase reaction.

A number of important observations can be made from this study. First, because the PVP leaves the circulation rather slowly, one may assume that the phagocytic cells have a relatively low affinity for PVP. Second, although the stimuli used are known to affect phagocytosis per se, it is by no means certain that the PVP is actually being cleared by phagocytosis. If the PVP exists in body fluid free of association with other macromolecules, then enhanced uptake may be a secondary phenomenon due to increased endocytic activity of the cell membranes with which it is associated. Alternatively, if PVP is associated with one or more plasma constituents, then enhanced uptake could reflect a change in turnover of the components, or it may simply be that PVP forms a complex with the substances administered (i.e., the HSA or fibrinogen) to enhance phagocytic uptake.

It may be concluded that uptake of PVP into reticuloendothelial cells in the body is most probably by a low affinity pinocytic mechanism. However, in situations in which phagocytosis is already active the PVP will be able to enter the cells by this mechanism also.

FUNCTIONAL CONSEQUENCES OF PVP UPTAKE INTO BODY TISSUES

Because PVP may be sequestered into the reticuloendothelial cells of the body, concern has been expressed regarding the effect this may have on the functioning of such tissues. Three aspects of this have come under scrutiny, mainly in relation to possible impairment of the homeostatic defense mechanisms of the body:

1. direct effect of PVP on the RES
2. PVP shock reactions
3. immunological effects of PVP

EFFECT OF PVP ON THE RES

Many polymers, including dextran (Farrows and Ricketts, 1971) and gelatin (Filkins and Di Luzio, 1966) inhibit reticuloendothelial activity. Caillard et al. (1976) reviewed the interactions of macromolecules with the RES. The reports about the effects of PVP on RES function are inconsistent. A number of papers (Schubert et al., 1951; Stern, 1952; Maruyama, 1960) showed that a reduction or blockade of RES activity could be produced by prior administration of PVP, but there are also in the literature a number of observations which show it to be totally without effect on RES function.

Experiments by Weikel and Lusky (1956) showed that intravenous infusion into rabbits of 10 ml/kg of a 3.5% solution of PVP (MW 40,000) weekly for 12 weeks (a total of 8.4 g to a 2-kg rabbit) reduced the phagocytic uptake of colloidal ^{32}P-chromic phosphate into the RES by 58% compared with untreated animals, affecting primarily uptake by the spleen. Since distribution studies using PVP have not shown it to be concentrated to any great extent by the spleen in particular, until large amounts are administered, it is possible that this phenomenon is associated only with the use of large quantities of PVP. The experiments of Smith (1959) would tend to confirm this suggestion. He gave 0.5 ml of 3.5% PVP (MW 40,000) to mice by intraperitoneal injection for 14 days (a total of about 0.25 g, or 10 g/kg) and produced a similar RES blockade, which inhibited the early clearance of Pasteurella pestis from the lungs and delayed the dissemination of Pasteurella pestis to other tissues.

In general agreement with these findings are the experiments of Alzetta (1967) and of Toyama (1965). Alzetta (1967) reported that PVP with a molecular weight of 25,000 decreased the disappearance time of horse albumin from the circulation of rabbits, while Toyama (1965), also using the rabbit, but PVP with the much larger MW of 700,000, showed a number of effects on the RES that indicated blockade. The PVP was administered intravenously (10–20 ml of a 1% solution) daily for 3 months. The author reported moderate swelling of liver and bone marrow RE cells but no other significant histological changes. A "severe" lymphocytopenia and a "mild" anemia were reported. RE cells in the spleen were reported to be ". . . markedly swollen with giant cell formation and complete disappearance of lymph follicles being accompanied with extensive fibrotic change. . . ." Also reported was a ". . . striking effect. . . in lymph nodes whose reticulum cells in the follicles were blocked up with heavy PVP deposition with poor lymphocyte production. . . ." After the period of PVP injections, the animals were given two intravenous doses of egg albumin (separated by 48 hours). Serum precipitin tests showed that there was a "mild" increase of serum-antibody titer in the PVP-treated animals.

Other work has shown that PVP is without demonstrable effect on RES activity both in vivo and in vitro. For example, Gans et al. (1967, 1968) showed that if rat livers were perfused in vitro with 5% glucose solution containing 3% PVP over a 30- to 60-minute period, there was no change in the uptake of colloidal gold or of bovine serum albumin. Similarly, PVP has been shown in vitro to be without effect on phagocytosis of Staphylococcus aureas by guinea pig neutrophils (Grzybek-Hryncewicz and Podolska, 1968). Using PVP with a reported MW of 355,200, phagocytosis was neither stimulated nor inhibited, but it did cause a decrease in the opsonizing properties of guinea pig serum and a reduction in serum complement titer. A similar effect was reported with polyvinyl alcohol with a MW of 32,560, but not with polyvinyl alcohol with a MW of 60,280 or 89,760, nor with Ficoll (MW 405,000) or dextran (MW 60,000–70,000). In vivo, Gelis et al. (1976) have shown that a single intravenous injection of PVP (8 ml/kg of 0.1%) did not affect the phagocytic uptake of colloidal gold in the rat. Neither did an identical dose of dextran. A modified fluid gelatin preparation decreased the index between 3 to 6 hours after administration.

Moody-Jones and Karran (1985) showed that divided doses up to 20 g/kg of PVP (MW 40,000) over a four-month period did not affect liver uptake of 99 m-technetium, indicating no adverse effect on hepatic reticuloendothelial function.

Pinocytotic activity generally leads to an increase in lysosomal enzyme activity in actively endocytotic cells (Cohn and Ehrenreich, 1969). It would be expected therefore, that PVP would cause an increase in lysosomal enzyme activity in RES cells. This appears to be the case. Meijer and Willighagen (1961) and Meijer (1962) demonstrated that PVP (MWs 17,000, 50,000 and 640,000), administered intraperitoneally to mice, caused an increase in the activities of the lysosomal enzymes, acid phosphatase and beta-glucuronidase, in the liver and spleen. They also reported that PVP was stored in the Kuppfer cells in the liver and in the cytoplasm of the reticulum cells of the spleen. In another study, Meijer and Willighagen (1963) found that the increase in liver acid phosphatase caused by the intraperitoneal administration of PVP (MW 12,600) in mice was not accompanied by a significant change in the nonlysosomal enzymes, a result consistent with enzyme studies in other reticuloendothelial cells following stimulation of pinocytosis.

Studies in rat hepatoma cells in culture have also shown that PVP K-25 and K-90 are without effect on alkaline phosphatase activity, although effects are induced by other macromolecules (Sorimachi and Yasumura, 1986).

One observation suggests that the extensive vacuolation induced by large amounts of PVP does not result in a decrease of intrapinolysosomal proteolysis. Work carried out by Roberts at al. (1976) using the rat visceral yolk sac showed that extensive vacuolation did not alter the digestion of bovine serum albumin.

PVP SHOCK REACTIONS

It has been shown (Halpern, 1956) that when PVP (MW 40,000) was administered intravenously to dogs in large amounts a shock reaction ensued similar to systemic anaphylaxis. This was not an immunologically mediated phenomenon but was due to histamine release from mast cells and was shown to occur in dogs but not rats. Since then, there have been a number of other studies concerned with the histamine-releasing properties of PVP. The clinical picture described by Halpern (1956) is that, within 2–3 minutes following injection, the skin becomes erythematous, followed by the appearance of urticarial wheals and general pruritus. Blood pressure begins to drop within 1–2 minutes of injection and the dog goes into shock. Recovery occurs within one to two hours, though in hepatectomized dogs, death is more likely (Adant, 1964). Other changes reported to accompany PVP-induced anaphylaxis include transient decreases in circulating platelets and granulocytes numbers and in fibrinogen level (Adant, 1964) and an increase in secretion of gastric juices (Halpern, 1956). Intravenous administration in dogs was found to have no effect on the motility of intestinal villi, mucus secretion or muscle tone, although decreased motility and increased mucus secretion were observed following injection directly into the mesenteric artery (Ihasz et al., 1966).

Plasma histamine has been shown to be elevated within

five minutes following intravenous PVP administration in dogs (Halpern, 1956; Yamasaki et al., 1969; Ruff et al., 1967). The early observations of the effect of PVP in dogs made by Halpern further support the role of histamine in PVP-induced anaphylaxis in this species. For example, intradermal administration of PVP in dogs results in a typical wheal and flare reaction, and prior administration of antihistamines (mepyramine, promethazine) blocks all the effects of PVP with the exception of increased gastric acid secretion. At the time of the early studies, the existence of two types of histamine receptors had not been elucidated.

Following PVP-induced anaphylaxis, dogs are refractory to a further challenge with PVP for up to six days (Halpern, 1956; Adant, 1964; Endo and Yamasaki,1969). A number of studies have been directed toward determining whether cross-tachyphylaxis exists between PVP and other histamine releasers. Studies have revealed no cross-tolerance between PVP and Witte's peptone (Ruff et al., 1967), sinomenine or compound 48/80 (Komoto, 1970). Some cross-tolerance has been observed by Komoto (1970) between PVP and Tween 10. Individual dogs have been identified that are resistant to the histamine-releasing activity of compound 48/80 and sinomenine. These dogs respond to PVP and are refractory to a secondary PVP challenge (Endo and Yamasaki, 1969). Dogs made refractory to PVP are still reactive to endotoxin-induced shock, which is not associated with a rise in plasma histamine (Corrado et al., 1964).

Tolerance to PVP is thought to be due to the depletion of the stores of histamine and the lack of cross-tolerance between PVP and other histamine releasers due to different sites of action. Nishiyama et al. (1957) measured the histamine content of liver, skin and muscle before and after PVP-induced anaphylaxis. Approximately 12% of the histamine was released from liver, 50% from skin and 10% from muscle. With compound 48/80-induced anaphylaxis, approximately 35% of liver histamine was released and 15% of skin histamine. The pattern of release associated with Tween 20 was similar to that of PVP. In the same study, PVP added to in vitro samples of liver and skin caused the same amount of histamine release from each. It appears that the differences in the sites of histamine release due to PVP and 48/80 may be due to differences in accessibility to the tissues. The only other species in which PVP-induced anaphylaxis appears to be adequately documented is in Charolais cows. During the course of the testing of a vaccine in Charolais cows in France, anaphylactoid reactions were observed which ranged from mild congestion to death in one case. A careful evaluation of the components of the vaccine allowed the isolation of PVP, which was present as a pharmaceutical aid, and this was considered to be the likely causative agent (J. Fois, personal communication, 1981; Schwartz and Herscowitz, 1982). Skin tests were done on reactive animals a year later with varying responses. The greatest responses were obtained in cows which had not been exposed to PVP the previous year. It is assumed that this was the primary encounter of these animals with PVP though cross-reactivity with another antigen; also, an unknown earlier exposure to PVP could not be eliminated (Schwartz and Herscowitz, 1982).

There is no conclusive evidence that PVP induces anaphylaxis in any other species. It has been reported to have no histamine-releasing activity in roosters (Lecomte and Beumariage, 1956) or in rats (Halpern, 1956). But, subplantar injection of PVP (MW 25,000) in the rat resulted in a transient edema similar to that induced by other irritants, including histamine and compound 48/80 (Bonta and De Vos, 1965; Bonta and De Vos, 1967). The authors suggest that the effect of PVP may be due to colloidal osmotic pressure, although no consideration was given to the possible direct histamine-releasing activity.

PVP-induced histamine release was examined in chopped lung preparations from guinea pigs, monkeys and humans. No release was observed from guinea pig lungs; release in the range of 0.1% to 1.5% was seen from monkey lung preparations; and slight release was seen from one of three human lung preparations tested. For purposes of comparison, PVP-induced release in this human specimen was 2% and compound 48/80-induced release was 10% (Nicholls, 1976).

There are no reported anaphylactoid reactions in humans which have been identified as being caused by PVP. One case was reported of a hospital worker who developed bronchial asthma as a result of exposure to a soybean powder which contained 2% PVP. In this patient, allergy to the soybean powder was confirmed, although the author stated that the direct histamine-releasing action of PVP might have played some role in the reaction (Peters, 1965). No support for such activity of PVP in this patient or in any other subjects was presented.

While it is generally accepted that PVP is a direct histamine-releaser in dogs, many unanswered questions remain. For example, some of the studies cited above suggest that more detailed studies of PVP-induced histamine release should be done in other species. In addition, the mechanism of action of PVP remains incompletely explored, and the possible contribution of such mediators as homocytotropic antibody or anaphylatoxin have not been thoroughly examined. In one study which examined the effects of various T-independent antigens, including PVP, on the activation of the alternate complement pathway (ACP), PVP was reported to activate the ACP, resulting in the generation of anaphylatoxin activity (Bitter-Suermann et al., 1975). Because of the manner in which the study was conducted, it was not possible to rule out a direct histamine-releasing activity of PVP.

IMMUNOLOGICAL EFFECTS OF PVP

The immunogenicity of chemical antigens differs very much in relation to their molecular weight and chemical and geometrical structure. In the case of linear molecules with repeating antigenic determinants, a T-dependent immunological response is commonly produced, although a low molecular weight may limit the ability to induce an immune response.

A number of studies have investigated the potential immunogenicity of PVP; one of the first was by Kerbel and Eidinger (1971). They examined the effect of various antigens in normal mice or mice that had previously been thymectomized (with or without additional treatment with antilymphocytic serum). They found that in normal animals PVP (grade not specified), when given intravenously, (0.1 μg/ animal) produced only a minimal amount of antibody. However, in animals that had been previously thymectomized and treated with ALS or just treated with ALS, there was a 4-fold increase in antibody titer during a 25-day period after PVP injection. They concluded that PVP was a thymus-independent antigen that produced a primary 19S immune response, which seems to be a reasonable assumption.

Subsequent work by Rotter and Trainen (1974) confirmed these results and went on to show that the potentiated response to PVP K-15 and PVP K-90 seen in thymectomized animals could be returned toward normal by thymus implantation or by a single injection of thymic cells.

Since then, Zimecki and Webb (1978) have used both in vivo (normal mice) and in vitro (whole spleen cell cultures) indicators of immune function and have shown that the extent of the primary 19S immunogenic reaction is dependent on the molecular weight (or size) of the PVP used, at least with respect to material with MWs of 360,000 and 40,000. However, they also showed that smaller molecules (MW 10,000) can be immunogenic, but only in the absence of significant numbers of T-cells, which appear to suppress the B-cell activation.

In in vitro studies Van Buskirk and Braley-Mullen (1987a) have shown that PVP is capable of activating either specific helper T-cells or suppressor T-cells, depending on the amount of PVP used in culture of mouse spleen cells. The same authors have also shown that PVP has similar effects in vivo (Van Buskirk and Braley-Mullen, 1987b).

With concern having been expressed about possible lymph

node involvement in PVP storage, a further study was initiated (Kaplan, 1984) to investigate the possible effect of locally administered PVP on one aspect of immune response, namely lymphocyte transformation. Three beagles were inoculated with 50 mg of PVP subcutaneously in 0.5 ml of phosphate-buffered saline into the front and rear right feet on each of the days 1, 3, 5, 8, 10, 12, 15, 17, 19, 22, 24, 26, 29 and 31. On day 16 the same dogs and three normal dogs were inoculated subcutaneously into each foot with 50 μg of turkey gamma-globulin (TGG) in Freund's adjuvant. On days 32 and 33 the right and left popliteal lymph nodes were removed from the dogs and the in vitro response of the lymph node cells to various stimuli was measured by their ability to incorporate tritiated thymidine (lymphocyte transformation).

It was found that there was considerable variability in the response between animals. However, no difference in lymphocyte transformation was observed between the lymphocytes from nodes removed from the right side (PVP inoculated) and the left side (control). It may be concluded that the effect of the nonspecific T-cell mitogens, concanavalin A and PHA was not inhibited by previous inoculation with PVP.

Furthermore, inoculation did not alter the proliferative response of lymph node cells to specific antigen challenge with PVP. A slight but statistically insignificant elevation of mitogen response was observed in the controls receiving turkey gamma-globulin alone.

Similar studies have been carried out in mice (Hoshi et al., 1986) in which PVP and other test substances, including thymus-dependent and thymus-independent antigens, were injected into the rear foot pad. The formation of germinal centers and lymph follicles in the draining popliteal lymph nodes was examined. PVP (MW 500,000) injection failed to induce a recognizable plasma cell reaction, germinal cell development or formation of new lymph follicles. The authors concluded, on the basis of the histological changes observed, that PVP is not very immunogenic.

SUMMARY OF MAIN FINDINGS

1. Evidence from a number of in vivo and in vitro studies in the rat shows that PVP enters cells mainly by fluid-phase pinocytosis.
2. PVP may also enter cells by phagocytosis where this process has been stimulated by the presence of other substances.
3. The effect of PVP on the reticuloendothelial system (RES) activity is variable and depends on the amount administered. When injected intravenously or intraperitoneally in large amounts, PVP reduces the phagocytic uptake of a number of agents by the RES, i.e., causes blockade of the RES. When given in smaller amounts, it has no demonstrable effect on RES activity either in vivo or in vitro.
4. Pinocytotic uptake of PVP into liver and spleen is accompanied by an increase in lysosomal enzyme activity in these organs.
5. A shock reaction has been reported following intravenous injection of large amounts of PVP in dogs which is similar to systemic anaphylaxis but which is not immunologically mediated. It is associated with an elevated plasma histamine level and has been ascribed to release of histamine from mast cells. Similar reactions have been reported in the cow, but not in rooster, rat or man.
6. Unlike other linear molecules with repeating antigenic determinants which induce a T-dependent immunological reaction, PVP induces a T-independent reaction.
7. PVP injected locally produces no change in lymphocyte transformation induced by nonspecific T-cell mitogens.

9

Toxicological Studies on PVP

ACUTE TOXICITY

Data on the acute toxicity of PVP of various MWs in mice, guinea pigs and rats have been tabulated in reports to the Joint FAO/WHO Expert Committee on Food Additives (WHO, 1980). LD_{50} values range from 12 g/kg intraperitoneally in the mouse to 100 g/kg orally in the rat (see Appendix 4). Such high doses of PVP when given orally have been shown to cause diarrhea in the dog, cat and rat, a fact which can be attributable to the nonabsorbable osmotic load. The dose required to produce this effect is species dependent but normally greater than 500 mg/kg.

Acute tolerance studies have additionally been carried out in rabbits, dogs and rhesus monkeys. No treatment-related histopathological changes were observed in these species.

In the rabbit study (Neumann et al., 1979), PVP with a MW of 50,000 was given by gavage in doses of 300, 900 or 2,700 mg/kg to groups of ten animals (5 male and 5 female). Controls received 20 ml of distilled water and three additional groups received hydroxypropylcellulose (HPC) at the same doses as PVP. Behavior, external appearance, food consumption, body weight, plasma enzymes and proteins, liver function and liver histopathology were all monitored. With the exception of a slight inhibition of body weight gain in the high-dose groups receiving either PVP or HPC, no treatment-related effects were reported.

In both the monkey and dog studies the PVP was given

intravenously, but even by this route no adverse effects were noted until the dose of PVP exceeded 10 g/kg. At this level a shock reaction was produced in both species which the authors concluded was related to the viscous hypertonic nature of the fluid injected. It was fatal in the monkey but not in the dog. Both experiments were performed by Hazelton Laboratories USA, for the National Cancer Institute (Zendzian and Teeters, 1970; Zendzian et al., 1981). One beagle of each sex was given 0, 1, 3 or 10 g/kg of PVP K-30 by intravenous infusion in a volume of 50 ml/kg. The animals were observed for a period of 28 days. Immediately after infusion of the highest dose the animals developed tremors and/or subconvulsive movements, defecated, salivated and showed depression and ptosis. They subsequently recovered, and apart from sporadic changes in transaminase and hematological values (which generally returned to normal within 2 days) remained normal throughout the rest of the study. At postmortem there were no gross or histological changes that could be related to treatment.

In the monkey experiments a very similar picture is evident. Three adult animals were used; one female received 5 g/kg, one male 10 g/kg and the remaining female, 0.9% saline. The higher dose, which was given as a 20% solution, killed the monkey after only half of the dose had been injected. Death was preceded by hyperactivity, bloody mucoid discharge from the mouth and nose, coma and loss of reflexes. At postmortem, although pulmonary congestion and edema were observed, this was not considered to be due to the chemical toxicity of PVP, but rather (as seen in the dog) to the viscous hypertonic nature of the 20% solution used. Because of these problems the 5 g/kg dose of PVP was given as a 10% solution. There were no immediate toxic effects of this injection and the animal survived without obvious ill effects until termination of the experiment 28 days later. Only minor transient changes in blood chemistry with elevated SGOT and SGPT levels were reported, and on histopathological examination no compound-related histological changes were seen.

SUBCHRONIC TOXICITY

Subacute toxicity studies have been conducted in rats, cats and dogs using PVP ranging in MW from 10,000 to 1,500,000 (Appendixes 5A and 5B). The PVP was administered at doses of 5–10 g/kg by gavage or 2.5–10% by weight in the diet over periods from 1.5 to 13 weeks. Many of the early studies had insufficient numbers of animals and/or were inadequately controlled and some of the findings are difficult to evaluate. In general, however, such studies showed no consistent treatment-related pathology, no abnormal histology, and blood and urine analyses which were performed were normal. Diarrhea, loose motions or fecal softening were common findings at the higher doses. The more recent controlled studies have confirmed this overall lack of effect.

Two rat and two dog studies are worthy of particular mention. In the first rat study (Shelanski, 1959a) PVP K-90 was administered at 2%, 5% or 10% by weight of the diet to groups of 50 (25 male and 25 female) Sherman-Wistar rats over a period of 90 days. Controls received normal diet. There was no significant difference in the weight gain curves, no consistent histological changes that could be attributable to PVP were noted, and special staining techniques for PVP were negative.

Similar findings were apparent in the second rat study (Kirsch et al., 1972). Sprague-Dawley rats received 2.5 or 5.0% PVP K-90 by weight in the diet over a 28-day period, after which it was reported that there were no toxic effects or pathological or histological findings that were related to PVP administration.

PVP K-90 was also used in both dog studies. In the first (Shelanski, 1959b) four groups of four Beagles (2 male and 2 female) received 2%, 5% or 10% by weight of PVP in the diet. Controls received diet alone. The animals receiving 10% PVP lost weight significantly, while those receiving control diet gained in weight. Despite this there was no consistent pathology associated with PVP administration, although special staining techniques for PVP revealed a positive reaction in mesenteric lymph nodes in all four animals receiving 10% PVP, in 3 out of 4 receiving 5% PVP, 1 out of 4 receiving 2% PVP, and 1 with a slight positive reaction in the controls. Shelanski (1959b) concluded that there was no significant effect caused by 2% PVP in the diet when compared with the controls.

In the second study (Kirsch et al., 1975) five groups of 8 Beagles (4 male and 4 female) received 2.5%, 5% or 10% PVP K-90, 10% cellulose, or diet alone over a 28-day period. Loose motions and diarrhea were observed periodically in the female animals receiving 2.5% PVP, but were seen consistently in all animals receiving 5% or 10% PVP. No other toxicity was reported, and at postmortem examination no changes were noted that could be related to the substances administered, although spleen weights were slightly raised in female animals receiving 10% PVP.

The two controlled rat studies both failed to produce significant changes at doses up to 10% of the dietary intake, which would indicate a no-effect level on the order of 5,000 mg/kg body weight. In the dog, 2% PVP in the diet was shown to be without effect, which again would indicate a no-effect level of about 5,000 mg/kg body weight.

PARENTERAL STUDIES

Local tissue tolerance

In studies carried out by BASF (1982, unpublished), local tissue tolerance to single or 5 repeated intramuscular injections of PVP was studied in female Sprague-Dawley rats using a similar protocol to the radioactive studies used by Cameron and Dunsire (1983a, 1983b, 1984). Eight groups of 9 animals were injected with 200 or 2,000 μg/animal in a volume of 12 μl of PVP K-12, K-17, K-30 or K-90 (2,000 μg of PVP K-90 was not injected because it was too viscous) or saline. A second set of eight groups of nine animals was injected with the same materials for 5 daily doses. Three animals were killed from each group 3, 14 and 45 days after dosing, and the site of administration (flexor digitorum profundus muscle of the left hind limb) examined histologically as well as the regional (popliteal iliac) lymph nodes.

After the single injections, there was evidence at 3 days of nonspecific tissue damage in some animals of all groups including controls, but by 14 and 45 days there was no evidence of any histological damage at the injection sites in any of the animals. After the 5 repeated injections, granulomatous tissue changes, typical of a foreign body reaction, were observed in 6 of 24 animals killed 3 days after treatment ended, 5 of 24 killed 14 days after treatment and in 2 of 24 killed 45 days after treatment. No correlation was observed with molecular weight of the PVP used, and it seems most likely that the reactions observed were due to the trauma of the injection procedures. No treatment-related effects were observed in the regional lymph nodes.

CHRONIC TOXICITY STUDIES

A number of studies have been carried out to look at the chronic toxicity of various grades of PVP (Appendixes 6A and 6B). Many of these have been part of an investigation of the possible carcinogenicity of this material. The earlier work was particularly concerned with potential dangers of intravenously and subcutaneously administered PVP in man, and this route was used in many tests in animals. While these have been referred to in relation to carcinogenicity, we have in this section confined our attention to PVP given orally. Three studies in rats and three in dogs are of interest.

The first study was by Shelanski (1957), who administered PVP K-30 to 50 male and 50 female Sherman-Wistar rats at 0, 1% or 10% by weight of the diet over a two-year period. The body weights of the top dose group were within 10% of the control weights throughout. Hematological parameters at 15, 18, 21 and 24 months were all within the normal range. Urine analysis at the same time intervals, including pH, sugar albumin content and density, showed no differences up to 15 months. Albumin was present in the urine of the top dose group at 18 months and in all groups including controls at 21 months. Although special attention was given to histological investigation of the lymphatic system, no gross or histopathological effects attributable to PVP were found in these or any other organs.

With the exception of frequent fluid stools, no other effects were observed. The author concluded that PVP administered to albino rats in quantities up to 10% by weight of the diet was harmless.

In the second study (BASF, 1978) 0,5% or 10% by weight PVP K-25 was administered in the diet over a 2-year period to Sprague-Dawley rats (50/sex/group). A second control group was fed 5% cellulose in the diet. No effect was found on food intake or body weight. No substance-related differences were found in hematological, clinical chemical or urine analysis parameters. No gross or histopathological effects were found related to treatment, and there was no evidence of PVP storage in mucous membranes of the duodenum or intestine or in the mesenteric lymph nodes.

A further study using PVP K-90 has been reported in which PVP was administered at 1%, 2.5% or 5% of the diet to groups of 75/sex/treated group and 125/sex/control group Sprague-Dawley rats (BASF 1980a). Controls received 5% cellulose. Groups of five animals of each sex were killed at weeks 26, 52 and 104, and similar groups were allowed 13-week recovery periods after cessation of dosing at these times, and were then killed in weeks 39, 65 and 117. The surviving male animals were killed at 129 weeks and the surviving females at 138 weeks. These timings were necessary in order to achieve a 70% mortality in the corresponding control groups as part of the carcinogenicity test. A total of 450 rats underwent treatment, and no treatment-related effects were observed on behavior, food and water intake, feces, body weight gain, hematology, ophthalmoscopy or auditory tests. No treatment-related effects were observed on organ weight or following macroscopic or histological examination of a wide range of tissues. There was no evidence of abnormal storage of PVP in the heart, liver, kidneys or lymph nodes using special staining techniques.

In the dog studies, PVP K-30 was used throughout. Two 1-year studies were carried out (Shelanski, 1958, and Wolven and Levenstein, 1957, quoted by Burnette, 1962) on a total of 32 dogs. Only brief summaries of the reports were published and the range of doses was not specified. These revealed no toxic effects, but one of the studies showed slight evidence of PVP assimilation from the gastrointestinal tract in the surrounding lymph glands at levels over 5%, while the other reported positive staining for PVP in the mesenteric lymph glands. However, since similar staining was also reported in the control animals, this latter observation must be viewed with some caution. In the second dog study, 16 beagles (2/sex/ group) received 0%, 2%, 5% or 10% PVP mixed with 10%, 8%, 5% and 0% Solka-Floc (a type of cellulose), respectively, in the diet over a two-year period. Overall there were no treatment-related effects, with the exception that swollen reticuloendothelial cells were found in lymph nodes in the group receiving 10% PVP. There was some evidence for similar findings at the 2% and 5% treatment levels, but to a lesser degree and less consistently. No treatment-related effects on food consumption, growth curves and hematology were observed. Postmortem gross and microscopic examination of tissues revealed no abnormalities. Since the swelling in the reticuloendothelial cells in the lymph nodes seen at the end of the one-year studies was similar in amount to that observed at the end of the two-year study, this indicated that no cumulative or extensive degenerative organ changes occurred in the second year period (Burnette, 1962).

REPRODUCTIVE TOXICITY STUDIES

Four tests have been performed, two in the rat and two in the rabbit, using PVP with four different MW distributions (see Appendix 9). In none of these was there evidence of an embryotoxic or a teratogenic action.

PVP K-25 was examined for prenatal toxicity in two groups of 25 female Sprague-Dawley rats. After mating overnight with untreated males, the presence of sperm in the vaginal smear was regarded as Day 0 of pregnancy. The PVP was administered in the diet (10% by weight) during the first 20 days of pregnancy to one group and the other group was given plain diet as control. On day 20 all the animals were killed and the fetuses removed. The body weight of the dams was recorded, also the number of resorption sites and the number of living and dead fetuses. The sex, weight and length of living fetuses and the placental weights were determined. Each fetus was then examined macroscopically for skeletal or organ malformations. Body weight gain in the treated dams was slightly reduced but there was no other evidence of toxicity in the dams. There were no malformations and no significant differences in the treated group in any of the parameters measured when compared with results from controls receiving normal diet (Zeller and Peh, 1976a).

In a similar study, PVP K-90 was given at 10% in the diet during the first 20 days of pregnancy to a group of 30 female Sprague-Dawley rats and a second group of 30 rats were given plain diet as controls. All animals were killed on day 20 for examination as above. The body weight gain in the treated dams was slightly reduced, but there was no other evidence of toxicity in the dams. No difference was observed in number of implants, resorptions or incidence of malformations in the treated groups compared with the controls (Zeller and Peh, 1976b).

Tests for embryotoxicity and teratogenicity were also conducted in groups of 11–12 Russian Chbb:HM rabbits, following artificial insemination (regarded as Day 0 of pregnancy). Doses of 50, 250 or 1250 mg/kg of PVP K-12 in saline were given daily by intravenous injection from days 6–18 of pregnancy. Controls were either untreated or given similar volumes of physiological saline intravenously. Caesarean section was carried out on all the animals on the 28th day after insemination. No adverse effects were observed on the dams given 50 or 250 mg/kg PVP. In the group injected with 1250 mg/kg PVP, food intake was reduced slightly and 8 of the 12 dams showed trembling, rapid breathing and convulsions lasting for approximately three minutes, following dosing on the second day only. The pregnancy rate was between 91–100% in all groups, with no treatment-related differences. There were no treatment-related effects on the number of corpora lutea, live implants, resorptions or abortions. There were no adverse effects on fetal weight or length or placental weight; and no differences in the incidence of variations or retardations. Only one malformed fetus was observed, and that was in the saline control group (Hofman and Peh, 1977).

A study has also been carried out using PVP with an average MW of 11,500, which was injected into the yolk sac of 9-day-old rabbit embryos (500 μg/embryo in 0.1 ml). Seventeen pregnant New Zealand White rabbits were used, and every yolk sac in one horn was injected with PVP and in the other horn with saline. Animals were killed on day 28. There was no difference in percentage of resorptions or malformations or in fetal weight or crown-rump length in the PVP-treated fetuses compared with controls injected with saline (Claussen and Breuer, 1975).

MUTAGENICITY STUDIES

Six investigations of the mutagenic potential of PVP have been carried out, all of which indicate that PVP is without mutagenic activity (Appendixes 8A and 8B).

In vitro tests

1. Ames test

PVP K-30 up to 5 mg/plate (Clairol Laboratories, 1978) or as 3.5% solution (Bruce, 1977) showed no activity with or without activation. PVP (up to 10 mg/plate) has also given negative results in the National Toxicology Program study (Zeiger et al., 1987).

2. Mouse lymphoma

PVP K-30 was tested in the L5178Y mouse lymphoma assay in a range of concentration from 1 mg/ml to 100 mg/ml with and without activation. No cell toxicity or mutagenicity was observed. EMS, MNNG and DMN were used as positive controls and showed expected activity (Kessler et al., 1980).

3. Cell transformation test

PVP K-30 was tested in a range of concentrations from 1–100 mg/ml for ability to transform Balb/C 3T3 cells. No activity was observed from PVP. MNNG was used as a positive control (Kessler et al., 1980).

In vivo tests

1. Dominant lethal test

Male mice were injected intraperitoneally once with 3.16 g/kg of PVP K-30. No effects on conception, number of implantations, percentage of live fetuses or mutagenic index were observed (Zeller and Engelhardt, 1977).

2. Bone marrow chromosomal aberration

Groups of five male and five female Chinese hamsters were injected with 3.16 g/kg i.p. of PVP K-30. At 4, 22 and 46 hours later, the animals were injected with colcemide and the bone marrow examined two hours later. 100 metaphases/animal were examined and no effect of PVP was detected (BASF, 1980b).

CARCINOGENICITY STUDIES
(see also Chronic Studies)

Oral studies (see also Appendixes 9B and 9D)

Three studies have been carried out in rats. In the first (Shelanski 1957), 0, 1% or 10% by weight of PVP K-30 was administered in the diet to groups of 50 male and 50 female rats for 2 years. Although the strain is not specified in the final report, it seems probable from other evidence that the strain was Sherman-Wistar. Half the rats were weanling (average weight 57 g) and the rest young adults (average weight 118 g males, 110 g females) at the start of treatment. Survival was satisfactory, with 26 of each sex at 1%, 24 males and 23 females at 10% and 27 males and 23 females in the control group surviving at 2 years. No toxicity was observed during the study except for 2–3% reduction of body weight at the top dose level; however, this dose exceeds the top dose currently recommended by the United States FDA of 5% of a test chemical in the diet, so it can be regarded as acceptable. No treatment-related abnormalities were detected in hematological examinations carried out in groups of 10 rats at 15, 18, 21 and 24 months. One male and one female per group were killed for autopsy at 18 months and 5 males and 5 females per group autopsied at 24 months. Viscera were examined grossly with particular attention to the lymphatic system, but only a limited number of organs (9 in all) were weighed and subjected to histopathological examination. There was no evidence of histopathological damage and no evidence of carcinogenicity.

In the second study (BASF 1978), 0, 5% or 10% by weight of PVP K-25 or 5% cellulose as control were administered in the diet to groups of 50 male and 50 female Sprague-Dawley rats for a period of 2 years. The rats had an average weight of 100 g at the start of treatment. Two control groups were used, one given just plain diet and the other given 5% cellulose in the diet. No dose-related toxicity was observed during treatment, and at term the numbers of animals surviving and available for examination were 5% PVP males 35, females 26; 10% PVP males 38, females 31; 5% cellulose males 34, females 31; absolute control males 40, females 38. Hematological examination was carried out in 10 rats per sex per group at 2, 4, 10, 16 and 24 months without evidence of toxicity. All surviving animals were necropsied at termination, and histological examination of the major tissues revealed no evidence of PVP storage or pathology, and the incidences of benign and malignant tumors were within normal limits in all groups.

In the third study (BASF 1980a) the higher molecular weight PVP K-90 was fed at 1%, 2.5% or 5% of the diet to groups of 75 male and 75 female Sprague-Dawley rats or 5% cellulose as control to 125 males and 125 females. Rats were 37–39 days of age (101–119 g) at start of treatment. Hematological examination was carried out in all surviving animals at the end of the study only. There was no evidence of dose-related toxicity. A total of 30 rats/sex/group were killed for examination at 26, 52 or 102 weeks. The remaining rats were killed when survival was 70% in the control groups, i.e., at 129 weeks for males and 138 weeks for females. At that time the numbers surviving in the various groups were 1% PVP 9 males, 25 females; 2.5% PVP 10 males, 32 females; 5% PVP 16 males, 27 females; controls 36 males, 37 females. Eleven major organs were weighed in all animals at autopsy and essentially a full range of organs (33 in all) subjected to histopathological examination. No evidence of carcinogenicity was found. The tumor incidence in the highest dose group did not show any influence of the test compound in respect to the nature, localization, number of affected animals, time of occurrence or ratio of benign to malignant.

Parenteral studies

Danneberg (quoted by Shubik and Hartwell, 1957) administered PVP "200 mg daily (total dose 200 g)" (molecular weight not stated) intravenously to 52 hybrid rats. The duration of the experiment was 32 months. There were 40 survivors at one year. There were tumors attributed to the PVP. Only one spontaneous uterine carcinoma after 12 months was reported. Lusky and Nelson (1957) published an abstract of a study in which 30 rats received subcutaneous injections of PVP (6%) at a dose of 1 ml/week, the molecular weight not identified. In addition, similar experiments were done with carboxymethylcellulose (CMC), 2%; Tween 60, 6%; and dextran, 6%. The rats were evaluated after 73 weeks for injection site tumors. Each group consisted of 30 rats consisting of 10 male and 10 female Osborne-Mendel rats and 10 male Bethesda black rats. Injection site tumors were identified in the CMC (43% incidence), PVP (43% incidence), and Tween 60 (17% incidence) groups. There were no sex differences and incidence tended to be somewhat higher in the Bethesda black rats. The authors identified the injection site tumors as ". . . fibrosarcomas of moderate histological malignancy." A thorough survey of the post-1957 literature failed to reveal a full paper by these authors. The route of administration in their experiments is similar to that which resulted clinically in granulomas and cutaneous storage effects (which were described earlier). It will also be recalled that such a phenomenon has also been mistaken for a malignant tumor.

HUEPER STUDIES ON PVP

Extensive early studies on the carcinogenicity of PVP were conducted by Hueper (1957, 1959, 1961). In the first study (Hueper, 1957), four forms of PVP were used, one of which was supplied by GAF (designated PVP I) and three by Schenley Laboratories (designated PVP II, PVP III and PVP IV). None were used for medicinal purposes. PVP I was identified as K-20 (MW 20,000; range 5,000–40,000), PVP II as K-22 (MW 22,000; range 5,000–40,000), PVP III as K-30 (MW 50,000; range 10,000–80,000) and PVP IV K-62 (MW 300,000; range 100,000–1,000,000). "The four PVP's were implanted in powdered form into the subcutaneous tissue at the nape of the neck and into the abdominal cavity. The individual dose was 200 mg for mice and 500 mg for rats." Twenty-five male and 25 female C57 black mice and 30 female Bethesda black rats were used for each route. In another series of experiments, each of the four PVPs was injected intravenously (femoral vein) to a group of 15 female rats at a dose of 2.5 ml of a 7% solution weekly for 8 weeks. All survivors were sacrificed at the end of 23 months from the start of the experiment. Hueper reported "tumors of the lymphoid and reticuloendothelial tissues, carcinomas of the uterus, skin, ovary and breast, and various benign tumors of some of these organs. The sites of the tumors were closely related to the sites at which PVP was retained . . ." He described 3 waves of tumors: reticuloendothelial sarcomas, Kupffer cell sarcomas and various carcinomas. For purposes of this discussion, lymphosarcomas, reticulum cell sarcomas and Kupffer cell sarcomas will be considered as RES sarcomas. It is difficult to summarize the nature of the controls used for this experiment, but the author states: "There were two sets of controls for the test series of C57 black mice. The first group was 75 untreated C57 black mice. The second was 954 of the C57 black mice that were used in other long-term experiments in which these animals were challenged by various chemical agents by either cutaneous applications or intramuscular

Table 10. Tumor Incidence in Mice Given Subcutaneous or
Intraperitoneal Implants of PVP[a]

Route	Average MW of PVP	RES Sarcomas	Carcinomas
Subcutaneous	20,000	0/50	0/50
(200 mg)	22,000	3/50	0/50
	50,000	0/50	0/50
	300,000	1/50	0/50
Intraperitoneal	20,000	0/50	0/50
(200 mg)	22,000	1/50	0/50
	50,000	3/50	0/50
	300,000	0/50	0/50

[a]See Hueper (1957) and text.

injections (carbon adsorbates of raw and filtered waters . . .
Bergius coal, oils and tars . . . Fischer-Tropsch oils), in order
to elicit chemospecific cancerous responses or to stimulate
and precipitate the development of tumors from tissues pos-
sessing hereditary or congenital liabilities to such develop-
ments." The results obtained with the PVP implants are presented in Table 10; no carcinomas were reported. In a study
of rats in the same paper, the incidence of carcinomas in the
animals was described. It is assumed that if carcinomas were
observed in the mice, they would have been described.
Benign tumors were also described in the rat study. Only
one was mentioned in connection with the mouse studies.
The only information that was given on the "controls" was
that the incidence of tumors was ". . . 0.4 percent for controls
. . . [and] . . . such tumors occurred only in controlled mice
painted with carcinogenic oils" [whatever that may mean—
Authors].

Studies with rats are described in another part of the same
report (Hueper 1957). The "controls" for the rat study were
also confusing. According to Hueper (1956), there were three
types of controls. "The first group was normal, untreated
rats of the Bethesda black strain (80 rats). The second was
Bethesda black rats either parenterally injected with various
chemicals (sheep fat, chromite-ore roast, uranium oxide) or

Table 11. Tumor Incidence in Female Rats Given PVP[a]

Route	Average MW of PVP	RES Sarcomas	Carcinomas	Benign Tumors
Subcutaneous	20,000	7/30	0/30	3/30
(500 mg of	22,000	0/30	1/30	3/30
powder/animal)	50,000	9/30	0/30	5/30
	300,000	7/30	1/30	3/30
Intraperitoneal	20,000	7/30	4/30	3/30
(500 mg of	22,000	2/30	0/30	3/30
powder/animal)	50,000	5/30	0/30	5/30
	300,000	12/30	0/30	2/30
Intravenous	20,000	2/15	1/15	0/15
(175 mg/animal	22,000	0/15	1/15	1/15
weekly × 8 wks)	50,000	7/15	0/15	1/15
	300,000	2/15	1/15	0/15

[a]See Hueper (1956, 1957) and text.

exposed to the inhalation of metallic-nickel dust and chromite-ore dust and surviving in part for more than 21 months (233 rats: 178 females, 55 males). The third group was Bethesda black rats that have received parenteral implantations of various types of dextran (240 female rats). The entire control group of 553 rats consisted of 453 females and 100 males."

Apparently, these animals were not observed for the same length of time as the animals which had received PVP. It is also questionable whether they were observed during the time that the PVP-treated animals were observed. Not all the animals described as controls died during the observation period, so that the only animals on which there appears to be pathology data are those which died during the study. The results from the PVP-treated animals are in Table 11 and the results from the "control" animals are in Table 12.

Hueper (1957) concluded that the mouse and rat data suggest that ". . . parenterally introduced . . . [PVP] . . . (1) favors the development of tumors of the lymphoid and RES in mice and (2) seems to be causally related to the appear-

Table 12. Tumor Incidence in Rats Used as Controls for PVP Study[a]

Group	Sex	Months of Observation	Number Died	RES Sarcomas	Carcinomas
Untreated	M	9	2	0	0
	F	13	23[b]	1	0
Exposed to	M	21	44[c]	0	0
various agents	F	21	129[d]	11	0
Exposed to dextran implants	F	15	68	2	0

[a]See Hueper (1957) and text.
[b]0 died in 0–3 months, 2 died in 4–9 months.
[c]12 died in 0–3 months, 9 died in 4–9 months.
[d]11 died in 0–3 months, 17 died in 4–9 months.

ance of benign and malignant neoplasms of various organs and tissues in the rats." These studies were poorly designed and inadequately and inappropriately controlled. Nevertheless, the index of suspicion for a causal relationship between the implantation of PVP powder and the occurrence of RES sarcomas is suggested. However, the assumption of a causal relationship between PVP and the occurrence of various carcinomas (all of which are commonly known to be spontaneous tumors in rats) and benign tumors is without adequate scientific foundation.

A similar study was again reported by Hueper (1959). This report contains new data and data reported elsewhere (Hueper, 1957; Hueper, 1961). Data in this paper (Hueper, 1959) that do not appear in either of the other two papers include the results of studies in which PVP was administered intravenously as a 7% solution to Dutch rabbits. The forms of PVP tested appeared to be those used in the previous study (Hueper, 1957). The dosing data are given in a table in which it is indicated that a total dose of 22 g was administered in single doses of 2.5 g. The frequency of administration was not indicated. (It is also recognized that 22 is not evenly divisible by 2.5). The rabbits were observed for "up to 4 years." No tumors were reported.

In the same report (Hueper, 1959), data were given on the

administration of PVP with an average molecular weight of 10,000 to mice and rats as a subcutaneous or intraperitoneal implant of the powder. The dose for mice (C57 black) was 200 mg; the dose for rats (Bethesda black) was 500 mg. The group size for mice was 30 and for rats 20. The ". . . maximal observation period . . . was two years." No tumors were observed in the mice which received the subcutaneous implant; one lymphoma and one mesothelioma of the pericardium were seen in mice which received the intraperitoneal implant. Four RES sarcomas were reported in the rats receiving subcutaneous implants; three RES sarcomas were seen in the rats which received the intraperitoneal implants. In the IARC Monograph on PVP (WHO, 1979), the intraperitoneal implant of PVP is erroneously identified as an ". . . i.p. injection of a PVP with a molecular weight of 10,000 . . . ." The distinction between an i.p. injection and an i.p. implant is critical, since it appears that the RES sarcomas are derived from a foreign body tumorigenic mechanism (see discussion below). Intraperitoneal injections of a substance, presumably in solution, would not be expected to be a likely cause of foreign body tumors.

Based on similar studies, Hueper also concluded that polyvinyl alcohol and dextran produce RES sarcomas (Hueper, 1959).

In a third study, Hueper (1961) emphasized molecular size as a study criterion in experiments with rats and rabbits. The study also involved the use of concurrent controls. Four forms of PVP were utilized, the first two of which were specially prepared by BASF. The first was PVP K-17, having an average molecular weight of 10,000 and a range of 2,000 to 38,000 with the major fraction being between 5,000 and 15,000. The second form was PVP K-25, having an average molecular weight of 18,000 with the major fraction having a range between 15,000 and 30,000. These averages for K-17 and K-25 are different from those which had been reported previously. Two other forms of PVP were used, both having an average molecular weight of 50,000. One was made by GAF and the other by BASF. The materials were adminis-

Table 13. Tumor Incidence in Rats and Rabbits Given PVP as an Aqueous Solution[a]

Species	PVP Type	Total Amount Administered	RES Sarcomas	Carcinomas
Rat	K-17	2.0 g	3/35	2/35
	K-25	2.0 g	1/35	2/35
	50,000 (GAF)	9.0 g	2/20	0/20
	50,000 (BASF)	9.0 g	2/30	3/30
	Control	9.0 g	2/30	3/30
Rabbit	K-17	62.2 g	0/6	0/6
	K-25	62.2 g	0/6	0/6
	Control		0/2	0/2

[a]See Hueper (1961) and text.

tered intraperitoneally in concentrations of 20% or 25% in divided doses over a period of approximately 6 to 10 weeks. Maximum survival was 24 months for rats and 28 months for rabbits. The data from the studies are shown in Table 13. Hueper concluded that the ". . . various types of cancers observed in the four series of animals receiving PVPs were identical in location and in structure with those seen in the control rats. . . [and in] . . . none of the test series were cancers more frequent than in the control series." In this study, as in the previous two studies (Hueper, 1957, 1959), Hueper noted the RES storage of PVP. Noting the absence of tumors in one group of rats which received the 50,000 MW PVP, Hueper stated ". . . the total absence of cancers among the rats receiving a solution of [that form of] PVP produces the impression that the massive storage of this PVP had hindered somehow the development of spontaneous cancer," i.e., PVP had an apparent anticarcinogenic effect. He cited the work of Stern et al. (1956), which reported a lower total incidence and a higher number of tumor-free old females in PVP-treated mice born of PVP-treated mothers (C3H). Treated parents and treated offspring received 24 subcutaneous injections of PVP (molecular weight not specified).

Other studies (BASF, 1958, BASF, unpublished)[a] were conducted to evaluate the potential carcinogenicity of PVP. In one study, PVP K-17 and PVP K-25 were administered intraperitoneally as aqueous solutions to male and female Bethesda black rats (received from Hueper), 5 g/kg at monthly intervals for a total of three doses. Two control groups were utilized. One was a manipulation control where the rats received distilled water intraperitoneally. The other was untreated. The number of animals in each group were: PVP K-17, 21 males and 21 females; PVP K-25, 36 males and 36 females; manipulation controls, 36 males and 36 females; untreated controls, 52 males and 21 females. No sarcomas were identified in the PVP K-17 group; 3 were identified in the PVP K-25 males; 1 was identified in the manipulation control males; and none were identified in the untreated controls. There were no significant differences in the incidences of carcinomas and benign tumors among the treated and control groups (BASF, 1958).

In another study (BASF, unpublished), Sprague-Dawley rats received a single subcutaneous implantation of PVP K-30, 500 mg. The controls were untreated. The treatment group contained 25 animals and the control group 12 animals. The average survival time of the treated group was 20 months. No sarcomas were observed in either group, nor was there a significant difference between groups in overall tumor incidence.

DISCUSSION OF THE PARENTERAL CARCINOGENICITY STUDIES

The development of sarcomas from the presence of undissolved PVP is probably a form of foreign body tumorigenesis. Foreign body tumorigenesis is probably a multistage development process where the physical *presence* of the for-

---

[a]Two of the authors reviewed all the original data cited. The conclusion that PVP is neither carcinogenic nor tumorigenic is supported by the data.

eign body, rather than its chemical composition, is the important etiologic factor. Sarcomas which result from the subcutaneous implantation of foreign bodies arise at the implantation site, but are capable of infiltration and metastases (Carter and Roe, 1969; Johnson et al., 1970; Brand et al., 1975). Since the subcutaneous administration of PVP did not result in metastases, PVP appears to differ from other "foreign bodies." A comprehensive and informative review of foreign body tumorigenesis has been published (Brand et al., 1976).

One current reference book states that "PVP parenterally in mice and rats is an established carcinogen" (Gosselin et al. 1976). The statement is made even though none of the references cited were concerned with the carcinogenicity of PVP. Available scientific data do not support the conclusion that PVP is a carcinogen. The history of the assessment of the carcinogenicity of PVP presents an interesting case for a critique of the assessment process itself, and also for the manner in which older studies are interpreted.

Foreign body tumorigenesis notwithstanding, the data and opinions discussed thus far do not support or even encourage the perception that "PVP is an established carcinogen." Ashwood-Smith (1971) stated that "concern about the possible connection between PVP and cancer has led to greatly reduced use in clinical practice." It is more likely that the reduced use of PVP as a plasma expander was a result of concern about tissue storage phenomena rather than cancer. Also, the methods developed for the treatment of hypovolemic shock by the U.S. military during the Vietnam war, and subsequently used in civilian practice, were directed away from the use of plasma expanders.

ANTICARCINOGENIC ACTIVITY OF PVP

The apparent anticarcinogenic effect of PVP has been described in a number of reports (Hueper, 1961; Stern et al., 1956). Chevallier et al. (1961a) reported that the carcinogenicity of 3,4-benzopyrene administered intramuscularly in the rat is decreased when the carcinogen is administered in a solution containing 40% PVP. In a subsequent study (Chevallier et al., 1961b) the authors tested the hypothesis that PVP was binding benzopyrene competitively with local protein such that the benzopyrene was eliminated with the PVP rather than remaining at the site of injection. The hypothesis was confirmed by studies of the special characteristics or organic extracts of the injection sites. Pamukcu et al. (1977) studied the effects of dietary PVP on the carcinogenicity of dietary bracken fern, a bladder and gastrointestinal carcinogen. The study was done in rats. PVP in the diet at a concentration of 50 mg/g caused a significant reduction in the percentage of animals with bladder tumors but not with gastrointestinal tumors. The study of Stern et al. (1956) on the apparent PVP-related reduction in the incidence of spontaneous mouse mammary cancer was cited above. Takeuchi (1966) reported on an experiment in which 1 ml of a 1% PVP K-30 solution was injected subcutaneously into a mouse, followed by inoculation with solid Ehrlich ascites tumor into the same site. Tumors were excised and weighed 8 days later. Animals treated with PVP showed a statistically significant 17% reduction in tumor weight. Hartveit (1967) reported that 5% PVP (MW 25,000) potentiates immune lysis of Ehrlich and Bergen A-4 ascites carcinoma cells.

The data on the potential carcinogenicity of PVP fail to demonstrate tumorigenic or carcinogenic effects. The local administration of PVP results in the development of sarcomas (a "foreign body" type reaction) but no metastases. A summary of carcinogenic studies is presented in Table 14.

Table 14. Summary of PVP Carcinogenesis Studies

Author	Year	Animal	Route	Time Span	Results and Remarks
Stern et al.	1956	Mouse Also first generation	Subcutaneous	500 + days	Negative: fewer tumors than in controls in treated mice and their offspring (fem.). Also, longer induction time to development of tumors in offspring.
Hueper	1957	Rats and mice Rats	Subcutaneous or IP as powder IV as 7% sol.	2 years	Borderline: mice Positive: rats
Hueper	1959	Mice and rats Rabbits	4–5 (not oral) IV	2 years	Borderline to negative in mice Positive – rats Negative – rabbits
Lindner	1960	Rats	IP	18 months	Negative
Hueper	1961	Rats and rabbits	IP	Rats – 2 yrs. Rabbits – 28 mos.	Negative: in both rats and rabbits. Actually fewer tumors in treated rats than in control rats (cf. Stern, above).
Burnette	1962	Rats	Up to 10% orally	2 years	Negative in all 4 studies (1 rats, 3 dogs)
		Dogs	Up to 10%	2 years (1 2-yr. study and 2 1-yr. studies)	
BASF (unpublished)	1967–69	Rats	Up to 10%	2 years	Negative
BASF (unpublished)		Rats	IP as 25% sol. 3X	Lifetime	No treatment-related effects
BASF (unpublished)		Rats	Subcut.; 0.5 gm PVP pdr.-1X	Lifetime	Negative

SUMMARY OF MAIN FINDINGS

1. Studies in rodents, dogs and primates have shown that PVP is a substance with a very low acute toxicity. It is essentially impossible to kill animals by administration of PVP except by gross osmotic imbalance. Thus the LD_{50} of PVP orally is reported to exceed 100 g/kg and to be around 10 g/kg or more intravenously or intraperitoneally. The only evidence of toxicity following oral administration is production of diarrhea with doses exceeding 0.5 g/kg due to the nonabsorbed osmotic load in the gut lumen.

2. Repeat-dose oral studies in rodents and dogs have shown that apart from diarrhea at high doses, related to the bulk purgative actions of PVP, there is no evidence of any toxicity as judged by clinical chemistry, hematology or histopathology. Occasional reductions in weight gain have been observed with 10% PVP in the diet, but this is probably related to reduced food intake and diarrhea. There is some evidence of minimal absorption as judged by the appearance of cell inclusions in the mesenteric lymph nodes in one dog study, but this is probably not of any toxicological significance. The no-adverse-effect level in subchronic studies in rodents and dogs is around 5 g/kg/day.

3. In two well conducted chronic studies using PVP K-25 and PVP K-90 at dose levels in the diet up to 10% and 5%, respectively, over periods of 2 years, there was no evidence of any substance-related toxicity in clinical chemistry, hematology, urine analysis or histopathology. There was no evidence of any carcinogenic effect or evidence of PVP storage in any organ. In less well reported chronic oral dog studies in which PVP K-30 was administered in the diet for 1–2 years, there was no evidence of treatment-related effects other than some evidence of PVP storage in the regional lymph nodes. There was no evidence of any cumulative damage over the 2 years.

4. After parenteral administration, PVP is well tolerated, with little evidence of local damage after single or 5 injections. Numerous studies on the chronic effects of injection of PVP in rats have given conflicting results, virtually all due to poorly designed and conducted studies. When well conducted studies have been carried out, there is no evidence of carcinogenic effect from repeated parenteral administration. Repeated injection of PVP, especially of larger molecular weight material, e.g., K-30 and over, will result in accumulation of PVP in the tissues. This is particularly so if the PVP is injected into poorly perfused sites. This has been demonstrated in animals to lead to development of foreign body type sarcomas at the site of injection but with no metastases.

5. Teratogenicity studies on PVP given orally, intravenously and intra-amniotically in rats and/or rabbits have shown no evidence of embryotoxicity or teratogenicity.

6. In vitro and in vivo mutagenicity studies have shown that PVP does not have any genotoxic or clastogenic activity.

10

Overall Summary and Conclusions

CHEMISTRY AND USES

PVP, polyvinylpyrrolidone, the homopolymer of N-vinyl-2-pyrrolidone, possesses unique physical and chemical properties that make it useful in many pharmaceutical formulations, foods, cosmetics and toiletries as well as industrial applications.

PVP preparations are available in different molecular weight ranges, with means varying from a few thousand to over one million. The letter K and an appropriate number, related to the molecular weight, are used to designate the different PVPs. For example, K-12 has an average molecular weight of around 4,000, and K-90 has an average molecular weight of around 1,000,000; however, there are differences in the formulae used by different manufacturers to calculate $\bar{M}_v$-values, so that there may be substantial differences in the molecular weights quoted by different authors. Furthermore, PVP, with a specified K-value and average molecular weight, consists of a range of molecular sizes with a slightly skewed bell-shaped distribution around the average.

BIOLOGICAL ACTIVITY

Because of its unique chemical nature, PVP would be expected to be biologically inert, apart from exerting osmotic activity. Extensive animal and human data support the inertness and hence the safety of PVP.

ABSORPTION

The absorption of PVP from the gastrointestinal tract is very limited and proceeds primarily by fluid-phase pinocytosis.

DISTRIBUTION, STORAGE, AND EXCRETION

PVP may be stored at the injection site or at distant sites (cutaneous storage syndrome). The extent of localized storage is dependent upon the molecular weight and amount of PVP injected, and the frequency and site of injections. The disappearance of PVP from the bloodstream is inversely related to the molecular weight and involves an initial rapid removal of low molecular weight materials by the kidney and a prolonged slower removal of higher molecular weight material into lymph and tissues, primarily RES tissues, including liver, spleen, bone marrow, bone and kidney. Tissue storage disease in humans occurs only following parenteral administration of large amounts (at least 200 g) of the higher molecular weight PVPs. Orally administered PVP is eliminated almost totally in the feces; very little is found in the bile or urine.

METABOLISM

PVP is not metabolized in the body. The excretion of residual monomer or low molecular weight oligomer present in the PVP under investigation could erroneously be interpreted as metabolism of PVP.

PHARMACOLOGICAL EFFECTS

Under normal use PVP does not modify physiological function. However, parenteral administration of large amounts has been reported to cause histamine release in the dog and Charolais cow, and other cardiovascular phenomena secondary to osmotic imbalance.

TOXICOLOGY AND SAFETY

An extensive body of toxicological data in animals supports the biological inertness of PVP. The acute, subchronic, and chronic toxicity of orally administered PVP is extremely low, with the only effect observed being diarrhea at high doses due to the osmotic action of PVP acting as a bulk purgative. Occasional observations of minimal absorption with storage in mesenteric lymph nodes seem to be of no toxicological importance. PVP is neither a sensitizer nor an irritant. There are no reported adverse effects following oral administration in humans. The currently permitted FAO/WHO ADI of 0-50 mg/kg body weight for food use provides an adequate margin of safety. There would appear to be no reason to restrict its oral or topical pharmaceutical use or topical cosmetic use in any way. There have been no reports of adverse effects following its use intravenously as a plasma expander, even after the administration of very large amounts. The only toxicological problems have involved the repeated injection of large amounts of the higher molecular weight material into poorly perfused sites such as subcutaneously and into the breast. If the use of PVP in injectables for repeated use is restricted to PVP with a molecular weight less than K-18 in limited amounts (e.g., 50 mg/i.m. dose) and the injection sites are varied, and intramuscular or intravenous routes are used, then these problems should not occur. The repeated use of PVP in depot preparations, which could lead to excessive storage, is not to be recommended.

References

ADANT, M. (1954) Quelques effets de l'injection intra-veineuse de Polyvinyl – pyrrolidone. Arch. Int. Physiol., 62, 145–146

ALTEMEIER, W.A., SCHIFF, L., GALL, E.A., GIUSEFFI, J., FREIMAN, D., MINDRUM, G. & BRAUNSTEIN, H. (1954) Physiological and pathological effects of long-term polyvinylpyrrolidone retention. A.M.A. Arch. Surg., 69, 309–314

ALZETTA, A. (1967) Permanenze in cincolo di un antigene dopo blocco con polivinil-pivrolidone (PVP) del sistema reti-culoendoteliale nel coniglio. Riv. Ist. Sieroter. Ital., 42, 191–195

ANGERVALL, L. & BERNTSSON, S. (1961) Oral toxicity of polyvinylpyrrolidone products at low average molecular weight. J. Inst. Brewing, 67, 335–336

AMEILLE, J., PAGES, M.G., CAPRON, F., PROTEAU, J. & ROCHEMAURE, J. (1985) Pathologie repiratoire induite par l'inhalation de laque capillaire. Rev. Pneumol. Clin., 41, 325–330

AMMON, R. & DEPNER, E. (1957) Ausscheidung und Verhalten verschiedener Polyvinylpyrrolidon-Typen im Organismus. Z. Geszmte. Exp. Med., 128, 607–628

AMMON, R. & MULLER, W. (1949) Der einfluss hoher Peristongaben auf den kaninchenorganismus unter beson-derer berucksichtigung der speicherorgane. Dtsch. Med. Wochenschr., 74, 465–468

ASHWOOD-SMITH, M.J. (1971) Polyvinylpyrrolidone solutions in plasma expanders: potential carcinogens? Lancet, 1, 1304

BASF (1958) Vorläufiger Bericht über die Prufung von Kollidon 30, Typ K26 und Dextran auf etwaige cancerogene Wirkung. Unpublished report (V/406, V/408)

BASF (1960) Bericht über die Prüfung von Kollidon K12 und K25 auf etwaige cancerogene Wirkung. Unpublished report (VII/72-73)

BASF (1978) 2-Jahre-Fütterungsversuch mit Kollidon 25 an der Ratte. Unpublished report

BASF (1980a) Chronic oral toxicity of Kollidon-90 USP XIX VERS NR 77-244 in Sprague-Dawley rats -repeated dosage over 129/138 weeks. Unpublished report submitted to JECFA 1983.

BASF (1980b) Effect of Kollidon 30 on bone marrow chromosomal aberration in Chinese hamsters. Unpublished report submitted to WHO 1980

BASF (1982) unpublished

BAZEX, A., GERAUD, J., GUILHEM, A., DUPRE, A., RASCOL, A. & CANTALA, P. (1966) Maladie de Dupont et Lachapelle (Thesaurismose cutanee par polyvinylpyrrolidone. Arch. Belg. Dermatol. Syphiligr., 22, 227

BEAHON, S.J. & WOODLEY, J.F. (1984) The uptake of macromolecules in adult rat columnar epithelium and Peyer's patch tissue in vitro. Biochem. Soc. Trans., 12, 1088

BERESFORD, C.R., GOLDBERG, L. & SMITH, J.P. (1957) Local effects and mechanism of absorption of iron preparations administered intramuscularly. Br. J. Pharmacol., 12, 107-114

BERGMANN, M., FLANCE, I.J. & BLUMENTHAL, H.T. (1958) Thesaurosis following inhalation of hair spray. A clinical and experimental study. New Eng. J. Med., 258, 471-476

BERGMANN, M., FLANCE, I.J., CRUZ, P.T., KLAM, N., ARONSON, P.R., JOSH, R.A. & BLUMENTHAL, H.T. (1962) Thesaurosis due to inhalation of hair spray. Report of twelve new cases, including three autopsies. New Eng. J. Med., 266, 750–755

BERT, J.M., BALMES, J.L., CAYROL, B., BALI, J.P., PAGES, A. & BALDET, P. (1972) Observation de thesaurismose a la polyvinyl pyrrolidone (P.V.P.) Sem. Hop. Paris, 48, 1809–1816

BITTER-SUERMANN, D., HADDING, U., SCHORLEMMER, H.L., LIMBERT, M., DIERICH, M. & DUKOR, P. (1975) Activation by some T-independent antigens and B-cell mitogens of the alternative pathway of the complement system. J. Immunol., 115, 425–430

BLAINEY, J.D. (1968) The renal excretion of higher molecular weight substances. Curr. Probl. Clin. Biochem., 2, 85–100

BOJSEN-MOLLER, M., RESKE-NIELSEN, E., VETNER, M. & HANSEN, J.C. (1976) PVP-aflejringssygdommen. Acta. Path. Microbiol. Scand., 84, 397–405

BONTA, I.L. & DE VOS, C.J. (1965) The effect of estriol-16,17-dihemi-succinate on vascular permeability as evaluated in the rat paw oedema test. Acta. Endocrinol., 49, 403–411

BONTA, I.L. & DE VOS, C.J. (1967) Significance of the kinin system in rat paw oedemas and drug effects on it. Eur. J. Pharmacol., 1, 222–225

BORK, K. (1982) Pseudotumoren und weitere Arzneimittelreaktionen durch nicht deklarierte Zusatzstoffe. Deutsche Med. Wochschr., 107, 43–45

BOYD, R.D.H., HILL, J.R., HUMPHREYS, P.W., NORMAND, I.C.S., REGNOLDS, E.O.R. & STRANG, L.B. (1969) Permeability of lung capillaries to macromolecules in foetal and new-born lambs and sheep. J. Physiol., 201, 567–588

BRAND, K.G., BUOEN, L.C. & BRAND, I. (1975) Foreign body tumorigenesis induced by glass and smooth and rough plastic. Comparative study of preneoplastic events. J. Natl. Cancer Inst., 55, 319–322

BRAND, K.G., JOHNSON, K.H. & BUOEN, L.C. (1976) Foreign body tumorigenesis, CRC Crit. Rev. Toxicol., 4, 353–394

BRÄUTIGAM, H.H. & GLEISS, J., (1956) Elimination of low-molecular polyvinylpyrrolidone in children with healthy kidneys. Zeitschrift für die gesamte experimentelle Medizin, 126, 531–536.

BRUCE, R. (1977) Unpublished report on polyvinylpyrrolidone in the Ames test. Carried out in the Ontario Cancer Institute for Dr.G.N.Ege of the Department of Nuclear Medicine, Princess Margaret Hospital, Toronto. Personal communication to GAF.

BUBIS, J.J., COHEN, S., DINBAR, J., HIRSCHHORN, B., SZEINBERG, A. & WOLMAN, M. (1975) Storage of polyvinylpyrrolidone mimicking a congenital mucolipid storage disease in a patient with Munchausen's syndrome. Israel J. Med. Sci., 11, 999

BÜHLER, V. & KLODWIG, U. (1984) Characterizing the molecular weight of polyvinylpyrrolidone. Acta Pharm. Technol. 30, 317–324

BULL, J.P., RICKETTS, C., SQUIRE, J.R., MAYCOCK, W.d'A., SPOONER, S.J.L., MOLLISON, P.L. & PATERSON, J.C.S. (1949) Dextran as a plasma substitute. Lancet, 1, 134–143

BUNDESGESUNDHEITSAMT (1983) Bundesanzeiger, No. 123/83, July 7, p. 6666

BURNETTE, L.W. (1962) A review of the physiological properties of PVP. Proc. Scien. Sect. of Toilet Goods Association, 38, 1–4

CABANNE, F., CHAPUIS, J.L., DUPERRAT, B. & PUTELAT, R. (1966) L'infiltration cutanee par la polyvinylpyrrolidone. Ann Anat. Pathol., 11, 385–396

CABANNE, F., MICHIELS, R., DUSSERRE, P., BASTIEN, H. & JUSTRABO, E. (1969) La maladie polyvinylique. Ann. Anat. Pathol., 14, 419–439

CAILLARD, B., VERRET, J., BLETTERY, B., LAS-SAUNIERE, J.M. & BENKHADRA, A. (1976) Macromole-cules, the reticuloendothelial system and immunologic problems. Anesth. Anal. Rean., 33, 597–613

CAMBRIDGE, G.W. (1973) Inhalation toxicity studies. Aerosol Age, 5, 32

CAMERON, B.D. & DUNSIRE, J.P. (1983a) The disposition of ¹⁴C polyvinylpyrrolidone in female rats following intra-muscular injection (PVP K17 and PVP K30). Unpublished report (No. 2505) by IRI for BASF.

CAMERON, B.D. & DUNSIRE, J.P. (1983b) The disposition of ¹⁴C polyvinylpyrrolidone in female rats following intra-muscular injection (PVP K12). Unpublished report (No. 2555) by IRI for BASF.

CAMERON, B.D. & DUNSIRE, J.P. (1984) The disposition of ¹⁴C-polyvinyl pyrrolidone in female rats following intramus-cular injection (PVP K30 excretion studies). Unpublished report (No. 2967) by IRI for BASF.

CAMPBELL, H., KANE, P.O., MUGGLETON, D.F. & OTTEWILL, I.G. (1954) Studies of excretion of polyvinylpyr-rolidone by normal human kidney. J. Clin. Path., 7, 252–260

CARTER, S.D., BRENNAN, F.M., GRACE, S.A. & ELSON, C.J. (1984) Clearance and organ localization of particles and soluble complexes in mice with circulating complexes. Immunology, 52, 761–769

CARTER, R.L. & ROE, F.J.C. (1969) Induction of sarcomas in rats by solid and fragmented polyethylene: experimental observations and clinical implications. Brit. J. Cancer, 23, 401–407

CAULET, T., ADNET, J.J. & HOPFNER, C. (1968) Thesauris-mose d'origine medicamenteuse. A propos de cinq observa-tions. Etude histochimique et ultrastructurale. Therapie, 23, 1049–1059

CHAPUIS, J.L., PUTELAT, R., DUPERRAT, B. & CABANNE, F. (1967) Cutaneous thesaurismosis caused by polyvinylpyrrolidone. Ann. Dermatol. Syphiligr., 94, 361–370

CHEN, W.Y.K., KEPES, J.J. & TEGLBJAERG, P.S. (1985) Intracellular mucoid changes in tumour cells of meningiomas: A manifestation of polyvinyl-pyrrolidone (PVP) effect on tissues with mesenchymal characteristics. J. Neuropathol. Exp. Neurol., 44, 606–616

CHEVALLIER, A., CHAMBRON, J. & MANUEL, S. (1961a) Influence de la presence de polyvinylpyrrolidone sur lacacerisation experimentale par le 3–4-benzo-pyrene. C. R. Seances Soc. Biol. Filiales, 155, 895–897

CHEVALLIER, A., MANUEL, S. & CHAMBRON, J. (1961b) Sur le mechanisme d'action de la polyvinylpyrrolidone dans la cancerisation experimentale par le 3–4-benzopyrene. C. R. Seances Biol. Filiales, 155, 916–918

CLAIROL LABORATORIES (1978) Ames test performed on six samples of hair spray resins. Unpublished report for GAF.

CLARKE, R.M. & HARDY, R.N. (1969a) The use of [125I] polyvinylpyrrolidone K-60 in the quantitative assessment of the uptake of macromolecular substances by the intestine of the young rat. J. Physiol., 204, 113–125

CLARKE, R.M. & HARDY, R.N. (1969b) An analysis of the mechanism of cessation of uptake of macromolecular substances by the intestine of the young rat ("closure"). J. Physiol., 204, 127–134

CLARKE, R.M. & HARDY, R.N. (1971) Factors influencing the uptake of [125I] polyvinylpyrrolidone by the intestine of the young rat. J. Physiol., 212, 801–817

CLAUSSEN, U. & BREUER, H.W. (1975) The teratogenic effects in rabbits of doxycycline, dissolved in polyvinylpyrrolidone, injected into the yolk sac. Teratology, 12, 297–302

COHN, Z.A. & EHRENREICH, B.A. (1969) The uptake, storage and intracellular hydrolysis of carbohydrates by macrophages. J. Exp. Med., 129, 201–226

COLOMB, D., PLAUCHU, M., LEUNG, T.K. & BIGON-

NET, A. (1970) Cutaneous thesaurismosis caused by polyvinylpyrrolidone (Dupont and Lachapelle disease). 3 new cases. Ann. Dermatol. Syphiligr. (Paris), 97, 249–258

COPENHAVER, J.W. & BIGELOW, M.H. (1949) Acetylene and carbon monoxide chemistry, Reinhold Publishing Co. (New York), p. 68

CORRADO, A.P., GARCIA LIMA, E. & ROTHSCHILD, A.M. (1964) Study of the mechanism of the cardiovascular shock produced by endotoxin of gram-negative bacteria: E. coli and S. typhii. Arch. Int. Pharmacodyn., 150, 462–482

COUINAUD, C., HERVE, J., BIOTOIS, C. & GIOAN, J. (1970) Thesaurismose a la polyvinylpyrrolidone revetant le masque d'une tumeur inflammatoire du grand epiploon. Sem. Hop. Paris, 46, 3079–3082

DEUTSCHER ARZNEIMITTEL-CODEX (DAC, 1979. Erg 1981) Monograph. "Losliches Polyvinylpyrrolidon," Govi-Verlag GMBH, Frankfurt, Deutscher Apotheker-Verlag, Stuttgart

DELBARRE, F., PAOLAGGI, J.B. & BASSET, F. (1964) Thesaurismose "artificiele" par polyvinyl-pyrrolidone. Presentation de deux cas. Bull. Mem. Soc. Med. Hop. Paris, 115, 243–258

DIGENIS, G.A., WELLS, D.A., ANSELL, J.M. & BLECHER, L. (1987) Disposition of ^{14}C povidone after oral administration to the rat. Food Chem. Toxicol., 25, 241–243

DOOLAN, P.D., SCHWARTZ, S.L., HAYES, J.R., MULLEN, J.C. & CUMMINGS, N.B. (1967) An evaluation of the nephrotoxicity of ethylenediaminetetraacetate and diethylenetriaminepentaacetate in the rat. Toxicol. Appl. Pharmacol., 10, 481–500

DORST, P.W., P.B. Report 4116, U.S. Department of Commerce, Office of Technical Services, 1945; Bibliogr. Tech. Rep., U.S. Department of Commerce, 1, 1945, p. 327

DUBOIS, R., GASSEE, J.P., DECOODT, P., STOUPEL, E., VERNIORY, A. & LAMBERT, P.P. (1975) The role of membrane parameters and of filtration pressure in the determination of the shape of the polyvinylpyrrolidone seiving curve. An in vitro and in vivo study. Contrib. Nephrol., 1, 9–20

DUBOIS, R. & STOUPEL, E. (1976) Permeability of artificial membranes to a pluridisperse solution of ^{125}I-polyvinyl-pyrrolidone Biophys. J., 16, 1427–1445

DUNCAN, R., PRATTEN, M.K., CABLE, H.C., RINGS-DORF, H. & LLOYD, J.B. (1981) Effect of molecular size of ^{125}I-labelled poly(vinylpyrrolidone) on its pinocytosis by rat visceral yolk sacs and rat peritoneal macrophages. Biochem. J., 196, 49–55

DUPONT, A. & LACHAPELLE, J.M. (1964) Dermite due a un depot medicamentaux au cours du traitement d'un diabete insipide. Bull. Soc. Franc. Dermatol. Syphiligr., 71, 508–509

DUVERNE, J., VOLLE, H., PERRET, M.J., LAGNIER, H. & BERNARD, J. (1971) Millaire pulmonaire: par thesaurismose on par allergie? J. Fr. Med. Chir. Thorac., 25, 69–73

ENDO, K. & YAMASAKI, H. (1969) Dogs refractory to compound 48/80 and sinomenine. Acta. Med. Okayama, 23, 589–592

FAIVRE, J., PECKER, J. & FERRAND, B. (1975) Syndrome de la fosse posterieure terminant l'evolution d'une histiocytose. X. Etude des lesions du systeme nerveux central. Association d'une thesaurismose a la polyvinyl-pyrrolidone. Sem. Hop. Paris, 51, 2229–2237

FAO (1986) Food and nutrition paper No. 37

FARROWS, S.P. & RICKETTS, C.R. (1971) Blocking of the reticuloendothelial cells by dextran. J. Pharm. Pharmacol., 23, 295–296

FARTASCH, M., ANTON-LAMPRECHT, I., FROSCH, P.J. & PETZOLDT, D. (1988) Polyvinylpyrrolidone dermatoses. Clinical aspects and ultrastructural morphology. Hautarzt, 39, 569–575

FEDERAL REGISTER (1971) 36, 16125

FELL, B.F., REGOECZI, E., CAMPBELL, R.M., & MACKIE, W.S. (1969) The permeability to ^{131}I-PVP of the normal and hypertrophied gastro-intestinal tract of sheep. Q. J. Exp. Physiol., 54, 141-155

FIKENTSCHER, H. & HERRLE, K. (1945) Mod. Plast., 23, 212-218

FILKINS, J.P. & DI LUZIO, N.P. (1966) Mechanism of gelatin inhibition of reticuloendothelial function. Proc. Soc. Exp. Biol. Med., 122, 177-180

FOIS, J (1981) Personal communication

FONCK-CUSSAC, Y., AUBLET-CUVELIER, J.L. & FONCK, J. (1970) Aspect ultra-structural de l'infiltration renale par la polyvinylpyrrolidone. Ann. Anat. Pathol., 15, 461-470

FOSSATI, P., DUPONT, A., CAPPOEN, J.P., DECOULX, M. & LINQUETTE, M. (1972) Forme hepatomegalique d'une thesaurimose a la polyvinylpyrrolidone apres traitement prolonge par la post-hypophyse retard. Rev. Franc. Endocrinol. Clin., 13, 57-62

FORDTRAN, J.S., RECTOR, F.C., EWTON, M.F., SOTER, N. & KINNEY, J. (1965) Permeability characteristics of the human small intestine. J. Clin. Invest., 44, 1935-1944

FRANK, H.B. & LEVY, G.B. (1953) Determination of molecular weight of polyvinylpyrrolidone. J. Polymer Sci., 10, 371-373

FROMMER, J. (1956) The pathogenesis of reticulo-endothelial foam cells. Effect of polyvinylpyrrolidone on the liver of the mouse. Am. J. Path., 32, 433-453

GAFFNEY, R.M. & CASLEY-SMITH, J.R. (1981) Excess plasma proteins as a cause of chronic inflammation and lymphoedema: biochemical estimations. J. Pathol., 133, 229-242

GALL, E.A., ALTEMEIER, W.A., SCHIFF, L., HAMILTON, D.L., BRAUNSTEIN, H., GIUSEFFI, J. & FREIMAN, D.G. (1953) Liver lesions following intravenous administration of polyvinylpyrrolidone (PVP). Am. J. Clin. Path., 23, 1187-1198

GANS, H., SUBRAMANIAN, V., LOWMAN, J.T. & TAN, B.H., (1967) Preservation of vascular patency as a function of reticuloendothelial clearance. II. Selectivity of phagocytosis for different clotting proteins of plasma. Surgery, 62, 698–703

GANS, H., SUBRAMANIAN, V. & TAN, B.H. (1968) Selective phagocytosis: a new concept in protein catabolism. Science, 159, 107–110

GARTNER, K., VOGEL, G., & ULBRICH, M. (1968) Studies on penetration of macromolecules (polyvinylpyrrolidone) through glomerular and post-glomerular capillaries of extravasal rotation of ^{131}I-albumin in the interstice of the kidney. Pflugers Archiv., 298, 303–321

GASSEE, J.P., ASKENASI, R. & LAMBERT, P.P. (1967) Etude de la permeabilite renale a des macromolecules. J. Urol. Nephrol. (Paris), 74, 375–391

GASSEE, J.P., VERNIORY, A. & LAMBERT, P.P. (1972) L'effet du clampage de l'aorte sur la pression de filtration glomerulaire efficace measuree a partir du tamisage du PVP. J. Urol. Nephrol. (Paris), 78, 1008–1011

GELIS, C., VIE, M.T. & MARIGNAN, R. (1976) Etude de l'influence de divers solutes de remplissage sur la captation de particules de dimension colloidale. Anaesth. Anal. Rean. (Paris), 33, 531–535

GENIAUX, M., ROLLIER, R., LAZRAK, B., M'SEFFAR, A. & IDRISSI, A. (1973) Cutaneous overload due to polyvinylpyrrolidone. Bull. Soc. Fr. Dermatol. Syphiligr., 79, 532–535

GILLE, J. & BRANDAU, H. (1975) Fremdkörpergranulation in der Brustdruse nach Injektion polyvinylpyrrolidonhaltigen Präparate. Geburtsh. u. Frauenheilk., 35, 799–801

GOSSELIN, R.E., HODGE, H.C., SMITH, R.P. & GLEASON, M.N. Clin. Tox of Comm. Prod. 4th ed., Williams & Wilkins, Baltimore, 1976, p. 245

GOWDY, J.M. & WAGSTAFF, M.S. (1972) Pulmonary infiltration due to aerosol thesaurosis. Arch. Environ. Health, 25, 101–108

GREENBLATT, D.J. & KOCH-WESER, J. (1976) Intramuscular injection of drugs. New Eng. J. Med., 295, 542–546

GRUNFELD, J.P., DE MONTERA, H., BERRY, J.P. & REVEILLAUD, R.J. (1968) A propos d'une observation the-saurismose par polyvinylpyrrolidone avec localisation renale predominante. J. Urol. Nephrol. (Paris), 74, 656–666

GRZYBEK-HRYNCEWICZ, K. and PODOLSKA, E. (1968) The influence of polyvinyl alcohol, polyvinylpyrrolidone, ficoll and dextran on phagocytosis. Arch. Immunol. Therap. Exp., 16, 702–708

HALPERN, B.N. (1956) CIBA Foundation Symposium on Histamine, pp.92.

HARANAKA, R. (1971) Intestinal absorption of polyvinyl-pyrrolidone. Nihon Univ. J. Med., 13, 129–146

HARDWICKE, J. (1972) Glomerular filtration of macromole-cules. Adv. Nephrol., 2, 61–74

HARDWICKE, J., HULME, B., JONES, J.H. and RICKETTS, C.R. (1968) Measurement of glomerular permeability to poly-disperse radioactively labelled macromolecules in normal rabbits. Clin. Sci., 34, 505–514

HARDY, R.N. (1965) Intestinal absorption of macromole-cules in the new-born pig. J. Physiol., 176, 19P-20P

HARDY, R.N. (1968) The acceleration by certain anions of the absorption of macromolecular substances from the small intestine of the new-born calf. J. Physiol., 194, 45P-46P

HARVEIT, F. (1967) In vitro potentiation of immune oncoly-sis by polyethylene glycol and by polyvinylpyrrolidone. J. Pathol. Bacteriol., 94, 200–204

HECHT, G & SCHOLTAN, W. (1959) Über die Auss-cheidung von Polyvinylpyrrolidon durch die normale Niere. Zeitschrift für die gesamte experimentelle Medizin., 130, 577–603

HECHT, G & WEESE, H. (1943) Periston, ein neuer Blutflus-sigkeitsersatz. Munch. med. Wochenschr., 90, 11–15

HEINRICH, H.C., GABBE, E.E., NASS, W.P. & BECKER, K. (1966) Untersuchungen zum Stoffwechselverhalten von ^{131}I-polyvinylpyrrolidon im menschlichen Körper. Klin. Wochenschr., 44, 488–493

HESPE, W., MEIER, A.M. & BLANKWATER, Y.J. (1977) Excretion and distribution studies in rats with two forms of ^{14}Carbon-labelled polyvinylpyrollidone with a relatively low mean molecular weight, after intravenous administration. Arzneimittelforschung, 27, 1158–1162

HIZAWA, K., OTSUKA, H., INABA, H., IZUMI, K. & NAKANISHI, S. (1984) Subcutaneous pseudosarcomatous polyvinylpyrrolidone granuloma. Am. J. Surg. Pathol., 8, 393–398

HOBER, R. & HOBER, J. (1937) Experiments on the absorption of organic solutes in the small intestine of rats. J. Cell. Comp. Physiol., 10, 401–422

HOFMAN, H.T. & PEH, J. (1977) Report on testing of Kollidon CE 5080 K12 (Compound No. XXV1/17-2) for prenatal toxicity in rabbits. Unpublished report for BASF, Gewerbehygeine und Toxikologie.

HOELSCHER, A.H. & ALTMANNSBERGER, M. (1982) Pseudotumor durch Injektionen von Procain-Polyvinylpyrrolidon. Dtsch. Med. Wochenschr., 107, 51–54

HONDA, K., MOTOKI, R., SAKUMA, H. & WATANABE, M. (1966) Complications following the use of plasma expander, especially polyvinylpyrrolidone. Intern. Surg., 45, 539–547

HORBACH, G.J.M.J., VAN LEEVWEN, R.E.W., SING HEIM YAP & VAN BEZOOIJEN, C.F.A. (1986) Changes in fluid-phase endocytosis in the rat with age and their relation to total albumin elimination. Mechanism of aging and development, 33, 305–312

HORT, E.V. & GASMAN, R.C. N-Vinyl monomers and polymers. In Encyclopedia of Chemical Technology. Ed. Kirk-Othmer, Vol. 23, 3rd. ed. John Wiley & Sons, Inc. (New York), 1983, pp. 960–979

HOSHI, H., KAMIYA, K., NAGATA, H., YOSHIDA, K. & AIJIMA, H. (1986) Formation of lymph follicles in draining lymph nodes after local injection of various antigenic substances in mice. Arch. Histol. Jap., 49, 25–37

HOUSLAY, M.D. and STANLEY, K.K. (1982) Dynamics of Biological Membranes, John Wiley & Sons (New York), p. 259

HUEPER, W.C. (1956) PVP, a carcinogenic agent for rats. Proc. Am. Assoc. Cancer Res., 2, 120

HUEPER, W.C. (1957) Experimental carcinogenicity studies in macromolecular chemicals, I. Neoplastic reactions in rats and mice after parenteral introduction of polyvinylpyrrolidone. Cancer, 10, 8–18

HUEPER, W.C. (1959) Carcinogenic studies on water-soluble and insoluble macromolecules. A.M.A. Arch. Path., 67, 589–617

HUEPER, W.C. (1961) Bioassay of polyvinylpyrrolidones with limited molecular weight range. J. Natl. Cancer Inst., 26, 229–237

HULME, B. (1975) Studies on glomerular permeability using inert polymers. Contr. Nephrol., 1, 3–8

HULME, B., & HARDWICKE, J. (1968) Human glomerular permeability to macromolecules in health and disease. Clin. Sci., 34, 515–534

IANNIELLO, R.M., COLONNESE, R. & MACHNICKI, N. (1987) Square-wave voltammetric determination of acetaldehyde in povidone. J. Assoc. Off. Anal. Chem., 70, 566–568

IHASZ, M., FUSY, J., KARIKA, G., KISS, I. & FEKETE, M. (1966) Anaphylactoid reaction and the intestinal mucous membrane. Kiserl. Orvostud., 18, 608–611

INDEST, H. (1978) Tissue levels in female rats as a function of time after intravenous administration. BASF Aktiengesellschaft. Unpublished report for BASF

INDEST, H. & BRODE, E. (1977) Excretion of PVP-K12 and PVP-K17 in female animals after intravenous application. BASF Aktiengesellschaft. Unpublished report for BASF

JARNUM, S (1961) The ^{131}I-polyvinylpyrrolidone (^{131}I-PVP) test in gastrointestinal protein loss. Scand. J. Clin. Lab. Invest., 13, 447–461

JOHNSON, K.H., BUOEN, L.C., BRAND, I. & BRAND,

K.G. (1970) Polymer tumorigenesis: clonal determination of histopathological characteristics during early preneoplasia; relationships to karyotype, mouse strain and sex. J. Natl. Cancer Inst., 44, 785–793

KAPLAN (1984) Evaluation of local effects of PVP on the canine immune response. Unpublished report to GAF

KERBEL, R.S. & EIDINGER, D.J. (1971) Variable effect of anti-lymphocyte serum on humoral antibody formation. Role of thymus dependency of antigen. J. Immunol., 106, 917–926

KESSLER, F.K., LASKIN, D.L., BORZELLECA, J.F. & CARCHMAN, R.A. (1980) Assessment of somatogenotoxicity of povidone-iodine using two in vitro assays. J. Environmental. Path. and Toxicol., 4, 327–335

KIRSCH, P., DATI, F., FREISBERG, K.O., BIRNSTIEL, H., MIREA, D. & ZELLER, H. (1972) Report on a study of the effects of Kollidon 90 when applied orally to rats over a 28 day period. BASF Gewerbehygeine und Toxikologie. Submitted to WHO by BASF

KIRSCH, P., DATI, F., FREISBERG, K.O., HEMPEL, K.J., MIREA, D., PEH, J., DECKARDT, K. & ZELLER, H. (1975) Report on a study of the effects of Kollidon 90 when applied orally to dogs over a 28 day period. BASF Gewerbehygeine und Toxikologie. Submitted to WHO by BASF

KOJIMA, M., TAKAHASHI, K. & HONDA, K. (1967) Morphological study on the effect of polyvinylpyrrolidone infusion upon the reticuloendothelial system. Tokohu J. Exptl. Med., 92, 27–54

KOMOTO, S. (1970) Histamine: Studies on the induced resistance and cross-tolerance to some histamine releasers. Okoyama Igahkai Zasski, 82, 147–158

KOSSARD, S., ECKER, R.I. & DICKEN, C.H. (1980) Povidone panniculitis. Arch. Dermatol., 116, 704–706

KUO, T. & HSUEH, S. (1984) Mucicarminophilic histiocytosis. A polyvinyl-pyrrolidone (PVP) storage disease simulating signet-ring cell carcinoma. Am. J. Surg. Pathol., 8, 419–428

LACHAPELLE, J.M. (1966) Thesaurismose cutanee par polyvinylpyrrolidone. Dermatologica (Basel), 132, 476–489

LACHAPELLE, J.M. & BOURLAND, A. (1967) Etude en microscopie electronique de la thesaurismose cutanee par polyvinylpyrrolidone. Bull. Soc. Fr. Dermatol. Syphiligr., 74, 538–544

LAMBERT, P.P., GASSEE, J.P., ASKENASI, R., FAF-CHAMPS, R., FICHEROULLE, P. & VERNIORY, A. (1970) La permeabilite glomerulaire aux macromolecules. Bull. Acad. Roy. Med. Belg., 10, 91–119

LAMBERT, P.P., GASSEE, J.P., VERNIORY, A. & FICHE-ROULLE, P. (1971) Measurement of the glomerular filtration pressure from sieving data for macromolecules. Pflügers. Arch., 329, 34–58

LAMBERT, P.P., VERNIORY, A. & GASSEE, J.P. (1972) Sieving equations and effective glomerular filtration pressure. Kidney Int., 2, 131–146

LECOMTE, J. & BEUMARIAGE, M.L. (1956) Liberation d'histamine et choc anaphylactique du coq. Comp. Rend. Soc. Biol., 150, 1028–1030

LE COULANT, P., TEXIER, L., LEURET, J. & GENIAUX, J. (1967) Iconographie d'un cas de thesaurismose par posthypophyse polyvidone. Bull. Soc. Fr. Dermatol. Syphiligr., 74, 675–676

LEDER, L.D. & LENNERT, K. (1972) Über iatrogene Lymphknotenveränderungen. Verh. d. Deutsch. Ges. f. Pathol., 56, 310–320

L'EPEE, P., LAZARINI, H.J., DOIGNON, J., DUTASTA, P. & N'DOKY, T. (1969) Cutaneous polyvinylpyrrolidone induced thesaurismosis complicating the treatment of post-traumatic diabetes insipidus. Med. Leg. Dommage. Corpor., 2, 184–186

LEUNG, T.K., LEUNG, J., PERROT, H. & THIVOLET, J. (1970) Ultrastructure study of skin infiltration by polyvinylpyrrolidone (substosan). Ann. Dermatol. Syphiligr. (Paris), 97, 567–572

LINDEMANN,B. & SOLOMON, A.K. (1962) Permeability of luminal surface of intestinal mucosal cells. J. Gen. Physiol., 45, 801–810

LINDNER, J. (1960) Tierexperimentalle Untersuchungen zum Problem der sogenannten Polymerkrebse. Verh. Dtsch. Ges. Pathologie, 44, 272–280

LOEFFLER, R.K. & SCUDDER, J. (1953) Excretion and distribution of polyvinylpyrrolidone in man as determined by use of radiocarbon tracer. Am. J. Clin. Path., 23, 311–321

LOEHRY, C.A., AXON, A.T.R., HILTON, P.J., HIDER, R.C. & CREAMER, B. (1970) Permeability of the small intestine to substances at different molecular weight. Gut, 11, 466–470

LOWSMA, H.B., JONES, R.A., PRENDERGAST, J.A. & BODERLOS, L.J. (1966a) Effects of respired polyvinylpyrrolidone aerosols in rats. Toxicol. Appl. Pharmacol., 8, 347

LOWSMA, H.B., JONES, R.A. & PRENDERGAST, J.A. (1966b) Effects of respired polyvinylpyrrolidone aerosols in rats. Toxicol. Appl. Pharmacol., 9, 571–582

LUSKY, L.M. & NELSON, A.A. (1957) Fibrosarcomas induced by multiple subcutaneous injections of carboxymethyl-cellulose (CMC), polyvinylpyrrolidone (PVP) and polyoxyethylene sorbitan mono-stearate (Tween 80). Abstract No. 1363. Fed. Proc., 16, 318

McCLANAHAN, J.S., LIN, Y.C. & DIGENIS, G.A. (1984) Disposition of N-vinyl-2-pyrrolidinone in the rat. Drug Chem. Toxicol., 7, 129–148

MAREK, H., KOCH, H. & SEIGE, K. (1969) Study of storage and elimination of tritium-labelled polyvinylpyrrolidone in the rat. Zeitschift fure die gesamte experimentalle Medizin, 150, 213–222

MARLOW, R. & FREEMAN, S.J. (1987) Differential effect of zinc on teratogen-induced inhibition of pinocytosis by cultured rat yolk sac. Life Sciences, 40, 1717–1723

MARUYAMA, Y. (1960) Urinary excretion of colloidal plasma

substitutes. Sapporo Igaku Zasski, 17, 256–269 (In Chemical Abstracts Vol 62, 9636, 1965)

MEIJER, A.E.F.G. (1962) The change in shape of the pH-activity curve of acid phosphatase in the liver and spleen of mice after intraperitoneal administration of macromolecular substances. Biochem. Pharmacol., 11, 125–134

MEIJER, A.E.F.G. & WILLIGHAGEN, R.G.J. (1961) Increased activity of acid phosphatase and beta-glucuronidase in the liver and spleen of mice after intraperitoneal administration of various macromolecular substances. Biochem. Pharmacol., 8, 389–397

MEIJER, A.E.F.G. & WILLIGHAGEN, R.G.J. (1963) The activity of glucose-6-phosphatase, adenosine triphosphatase, succinic dehydrogenase and acid phosphatase after dextran or polyvinylpyrrolidone uptake by liver in vivo. Biochem. Pharmacol., 12, 973–980

MERKLE, J., MIREA, D. and HILDEBRAND, B. (1983) Study of various PVP grades for their local tissue tolerance after 1 and 5 intramuscular injections. BASF Aktiengesellschaft. Translation of unpublished report

MEZA, R. & GARGALLO, L. (1977) Unperturbed dimensions of polyvinyl-pyrrolidone in pure solvents and in binary mixtures. Eur. Polymer J., 13, 235–239

MILLER, S.A. (1965) Acetylene, Its Properties, Manufacture and Uses. Vol. 2, Academic Press (New York), pp. 338–339

MOHN, G., (1960) Storage of polyvinylpyrrolidone in rat organs, shown by direct fluorescence microscopic determination. Acta Histochem., 9, 76–96

MOINADE, S., TERASSE, J. & VILATTE, A. (1977) A propos d'un cas de thesaurismose polyviscerale a la polyvinyl pyrrolidone. Ann. Med. Interne. (Paris), 128, 183–187

MOODY-JONES, D. & KARRAN, S.J. (1985) Studies of hepatic tolerance of polyvinylpyrrolidone in the rat. J. Hosp. Infect., 6 (Suppl.), 205–207

MORGENTHALER (1977) Excretion of PVP-K25 and Collidone VA 64 in comparison to that of PVP-K12 in male animals. Determination of amounts absorbed. BASF Aktienge-sellschaft. Unpublished report.

MÜLLER, W. (1946) Zur pathologischen Anatomie der alimentären Intoxikation. Deut. Med. Wochenschr., 71, 32

NELSON, A.A. & LUSKY, L.M. (1951) Pathological changes in rabbits from repeated intravenous injections of Periston (polyvinyl pyrrolidone) or dextran. Proc. Soc. Exp. Biol. Med., 76, 765–767

NEUMANN, A., LEUSCHNER, A., SCHWERTFEGER, W. & DONTENWILL, W. (1979) Study on the acute oral toxicity of PVP (MW 50,000) in rabbits. Unpublished report to BASF

NEVINS, M.A., STETCHEL, G.H., FISHMAN, S.I., SCHWARTZ, G. & ALLEN, A.C. (1965) Pulmonary granulomatosis. Two cases associated with inhalation of cosmetic aerosols. J. Am. Med. Assoc., 193, 266–271

NICHOLLS, P.J. (1976) Release of histamine from lung tissue in vitro by dimethylhydantoin-formaldehyde resin and polyvinylpyrrolidone. Brit. J. Ind. Med., 33, 127–129

NISHIYAMA, R., TASAKA, K. & IRINO, S. (1957) The sites of action of some histamine-releasing substances in the dog. Acta Med. Okayama, 11, 133–144

OETTEL, H. & VON SCHILLING, B. (1967) Report on testing of various collidones and dextran for storage in organs and potential carcinogenic effect in the rabbit. BASF Aktiengesellschaft. Unpublished report in German

OWEN, M.C., IMMELMAN, A. & GRIB, D. (1975) The elimination of albumin, polyvinylpyrrolidone and dextran from the circulation in sheep. J. S. Afr. Vet. Assoc., 46, 245–247

PAMUKCU, A.M., YALCINER, S. & BRYAN, G.T. (1977) Inhibition of carcinogenic effect of bracken fern (Pteridium aquilinum) by various chemicals. Cancer (Suppl.), 40, 2450–2454

PAPPENHEIMER, J.R., RENKIN, E.M. & BORRERO, L.M. (1951) Filtration, diffusion and molecular seiving through peripheral capillary membranes. A contribution to the pore theory of capillary permeability. Am. J. Physiol., 167, 13–46

PETERS, G.A. (1965) Bronchial asthma due to soybean allergy: report of a case with audiovisual documentation. Ann. Allergy, 23, 270–272

PLAUCHU, M., COLOMB, D., LEUNG, T.K., POUSSET, G. & BIGONNET, A. (1970) L'infiltration tissulaire par la polyvinylpyrrolidone. A propos de 3 nouveaux cas de thesaurismose cutanee apres emploi prolonge de posthypophse polyvidone (Maladie de Dupont et Lachapelle). Lyon Med., 223, 1133–1145

PRATTEN, M.K. & LLOYD, G.B. (1986) Pinocytosis and phagocytosis: the effect of size of a particulate substrate on its mode of capture by rat peritoneal macrophages cultured in vitro. Biochimica et Biophysica Acta, 881, 307–313

PRATTEN, M.K., WILLIAMS, K.E. & LLOYD, J.B. (1977) A quantitative study of pinocytosis and intracellular proteolysis in rat peritoneal macrophages. Biochem. J., 168, 365–372

PRINCIOTTO, J.V., RUBBACKY, E.P. & DARDIN, V.J. (1954) Two year feeding study in dogs with polyvinylpyrrolidone (Plasdone C). Unpublished report from Chemo Medical Consultants (USA) for GAF. Submitted to WHO by BASF 1954.

RASK-MADSEN, J. (1973) Seiving characteristics of inflamed rectal mucosa. Gut, 14, 4988–4989

RAVIN, H.A., SELIGMAN, A.M. & FINE, J. (1952) Polyvinylpyrrolidone as a plasma expander. Studies on its excretion, distribution and metabolism. New Eng. J. Med., 247, 921–929

REGOECZI, E. (1976) Labelled polyvinylpyrrolidone as an indicator of reticuloendothelial activity. Br. J. Exp. Path., 57, 431–442

REPPE, W. (1949) Acetylene Chemistry, PB report No. 18852-S, U.S. Department of Commerce, Charles A Meyer and Co. Inc., New York, pp 68–72

RESKE-NIELSEN, E., BOJSEN-MULLER, M., VETNER, M. & HANSEN, J.C. (1976) Polyvinylpyrrolidone-storage disease, light microscopical, ultrastructural and chemical verification. Acta. Path. Microbiol. Scand., 84, 397–405

RHEINHOLD, J.G., VON FRITJOG DRABBE, C.A.J., NEWTON, M. & THOMAS, J. (1952) Effects of dextran and of polyvinylpyrrolidone administration on liver function in man. Arch. Surg., 65, 706–713

RHOADS, J.E. (1952) Various plasma expanders in man. Ann. N. Y. Acad. Sci., 55, 522–525

RIMBAUD, P., MARTY, C., MAYNADIER, J. & GUILHOU, J.J. (1971) Dupont-Lachapelle disease. Cutaneous and visceral thesaurismosis due to polyvinylpyrrolidone. Bull. Soc. Fr. Dermatol. Syphiligr., 78, 242–244

ROBERTS, A.V.S., NICHOLLS, S.E., GRIFFITHS, P.A., WILLIAMS, K.E. & LLOYD, J.B. (1976) A quantitative study of pinocytosis and lysosome function in experimentally induced lysosomal storage. Biochem. J., 160, 621–629

ROBSON, M., ELIRAZ, A. & ROSENFELD, J.B. (1973) Determination of glomerular permeability in renal transplantation using polyvinylpyrrolidone. Israel J. Med. Sci., 9, 424–428

ROSE, C.E. & WHITE, R.D. (1945) PB Report 1308, U.S. Department of Commerce, Office of Technical Services; Bibliogr. Tech. Rep., U.S. Department of Commerce, 1, p. 223

ROTTER, V. & TRAINEN, N. (1974) Thymus cell population exerting a regulatory function in the immune response of mice to polyvinyl-pyrrolidone. Cell. Immunol., 13, 76–86

RUFF, F., SAINDELLE, A., DUTRIPON, E. & PARROT, J.L. (1967) Continuous automatic fluorometric evaluation of total blood histamine. Nature, 214, 279–281

RUFFER, W. (1955) Klinische Nachuntersuchungen zur Frage der Kollidon-Speicherung. Medizinische, 15, 539–541

SANNER, A., STRAUB, F. & TSCHANG, C.H. (1983) Chemistry, structure and properties of polyvinylpyrrolidone. In Proceedings of the International Symposium on Povidone. Ed. G. A. Digenis & J. Ansell. University of Kentucky, pp. 20–38

SCHEFFNER, D. (1955) Tolerance and side effects of various Kollidons administered by mouth and their behaviour in the gastrointestinal tract. Doctors' Thesis. University of Heidelberg.

SCHILLER, A. & TAUGNER, R. (1980) The renal handling of low molecular weight polyvinylpyrrolidone and inulin in rats. In Functional Ultrastructure of the kidney – Proceedings of International Symposium, 315–326

SCHOEN, H. (1949) Organveränderungen beim Säugling nach Zufuhr von Periston. Klin. Wochenschr., 27, 463–468

SCHOLTAN, W. (1951) Molekulargewichtsbestimmung von Polyvinylpyrrolidon mittels der Ultrazentrifuge. Makromol. Chem., 7, 209–235

SCHUBERT, R., SEYBOLD, G. & WUNDT, H. (1951) Der Einfluss von niedermolekularem Kollidon auf die Vitalfärbung in der Niere im RES. Deut. Z. Verdauungs, 11, 63–69

SCHWARTZ, S.L. (1981) Evaluation of the safety of povidone and crospovidone – a review article. Yakuzaigaku, 41, 205–217

SCHWARTZ, S.L. & HERSCOWITZ, H.B. (1982) Unpublished report to GAF.

SENAK, L. WU, C.S. & MALAWER, E.G. (1987) Size exclusion chromatography of poly(vinylpyrrolidone). II. Absolute molecular weight distribution by SEC/LALLS and SEC with universal calibration. J. Liq. Chromatogr., 10 (6), 1127–1150

SHALLOCK, Von G. (1943) Anatomische Untersuchungen uber das Schicksal von Blutersatzmitteln im Emptängerorganismus und der durch sie ausgelösten Reaktionen. Beitr. Path., 108, 405–451

SHELANSKI, M.V. (1953) PVP K-30 ^{14}C single dose excretion study. Unpublished report for GAF

SHELANSKI, M.V. (1957) Two year chronic oral toxicity study with PVP K-30 in rats. Unpublished report from the Industrial Biological Research and Testing Laboratories (USA) for GAF. Submitted to WHO by BASF

SHELANSKI, M.V. (1958) One year feeding study in dogs

with Plasdone C from the Industrial Biological Research and Testing Laboratories (USA). Unpublished report for GAF

SHELANSKI, M.V. (1959a) Ninety-day feeding study with PVP K-90 in rats. Unpublished report from the Industrial Biological Research and Testing Laboratories (USA) for GAF. Submitted to WHO by BASF

SHELANSKI, M.V. (1959b) Ninety-day feeding study with PVP K-90 in dogs. Unpublished report from the Industrial Biological Research and Testing Laboratories (USA) for GAF. Submitted to WHO by BASF

SHELANSKI, M.V. (1960) Metabolic tracer studies with radioactive PVP K-90. Unpublished report for GAF

SHELANSKI, A.A., SHELANSKI, M.V. & CANTOR, A. (1954) Polyvinylpyrrolidone (PVP) as a useful adjunct in cosmetics. J. Soc. Cosm. Chem., 5, 129–132

SHUBIK, E., & HARTWELL, J.L. (1957) U.S. Dept. Health, Education and Welfare, Entry 909, p. 301, suppl. 1.

SIBER, G.R., MAYER, R.J. & LEVIN, M.J. (1980) Increased gastrointestinal absorption of large molecules in patients after 5-fluorouracil therapy for metastatic colon carcinoma. Cancer Res. 40, 3430–3436

SLIFKIN, S.C. (1947) P.B. Report 78256, U.S. Department of Commerce, Office of Technical Services, 1947; Bibliogr. Tech Rep., U.S. Department of Commerce, 7, p. 616

SMITH, P.N. (1959) Pneumonic plague in mice; the effect of polyvinylpyrrolidone. Am. J. Hygiene, 69, 21–24

SORIMACHI, K. & YASUMURA, Y. (1986) Regulation of alkaline phosphatase activity in rat hepatoma cells. Effects of serum proteins, cycloheximide, actinomycin D, chloroquine, dinitrophenol and potassium cyanide. Biochimica et Biophysica Acta, 885, 272–281

SOUMERAI, S. (1978) Pseudotumours of the arm following injections of procaine polyvinylpyrrolidone. Report of two cases. J. Med. Soc. N. J., 75, 407–408

STAHL, C.R. & FRAUENFELDER, L.J. (1972) Excretion studies on Plasdone C. Unpublished report for GAF

STERN, K. (1952) Effect of polyvinylpyrrolidone on reticulo-endothelial storage. Proc. Soc. Exp. Biol. Med., 79, 618–623

STERN, K., SABET, L. & GLEASON, M. (1956) Effect of polyvinylpyrrolidone (PVP) on incidence of spontaneous mouse mammary cancer. Proc. Am. Assoc. Cancer Res., 2, 150

SUND, R.B. & SCHOW, J. (1964) The determination of absorption rates from rat muscles: an experimental approach to kinetic disposition. Acta Pharmacol. Toxicol. (Kbh), 21, 313–325

TAKEUCHI, J. (1966) Growth-promoting effect of acid muco-polysaccharides on Ehrlich ascites tumor. Cancer Res., 26, 797–802

TANAKA, M. (1971) Two autopsy cases with accumulation of low molecular weight polyvinylpyrrolidone (PVP). Naika, 28, 389–393

TEREAU, Y., VONNET, A., MATAR, A. & BROCHERION, G. (1978) Localisation mandibulaire d'une thesaurismose (polyvinylpyrrolidone). Rev. Stomatol. Chir. Maxillofac., 79, 91–97

THIVOLET, J., LEUNG, T.K., DUVERNE, J., LEUNG, J. & VOLLE, H. (1970) Ultrastructural morphology and histo-chemistry (acid phosphatase) of the cutaneous infiltration by polyvinylpyrrolidone. Br. J. Dermatol., 83, 661–669

TOKIWA, T. (1958) Polyvinylpyrrolidone. Sogo Igahu 15, 159–171

TOWERS, R.P. (1957) Lymph node changes due to polyvinyl-pyrrolidone. J. Clin. Path., 10, 175–177

TOYAMA, T. (1965) Studies on the function of reticulo-endothelial systems. II. Effects of the R.E.S. blocking with macro-molecular PVP on the lymphoid cell reproduction and the production of serum antibody. Acta. Med. Okayama, 19, 307–316

TRAENCKNER, K. (1954a) Das schicksal des Peristons im menschlichen Körper nach histologischen unter suchungen. Frankfurt Ztschr. Path., 65, 62–79

TRAENCKNER, K. (1954b) Zur Frage des Schicksals des Dextrans im menschlichen Körper nach histologischen Untersuchungen. Frankfurt Ztschr. Pathol., 65, 390–408

UFFENORDE, J., BRUNNER, F.X. & WUNSCH, P-H. (1984) Polyvinylpyrrolidone granuloma – differential diagnosis of rare soft tissue tumors in the area of the maxilla, HNO, 32 (12), 515–517

ULLMANS ENCYKLOPADIE DER TECHNISCHEN CHEMIE, 4, Auflage 1980 Band 19, p. 385

UPHAM, H.C., LOVELL, F.W., DETRICK, L.E., HIGHBY, D.H., DEBLEY, V. & HALEY, T.J. (1956) Tissue deposition of polyvinylpyrrolidone in normal and irradiated rabbits. Arch. int. Pharmacodynam., 106, 151–163

UNITED STATES PHARMACOPEIA, Twenty-first revision (1985) through Supplement 7, United States Pharmacopeial Convention Inc.

VAN BUSKIRK, A.M. & BRALEY-MULLEN, H. (1987a) In vitro activation of specific helper and suppressor T cells by the Type 2 antigen polyvinyl-pyrrolidone (PVP). J. Immunol., 139, 1400–1405

VAN BUSKIRK, A.M. & BRALEY-MULLEN, H. (1987b) Suppression of IgG memory responses by T cells associated with the Type 2 antigen polyvinyl-pyrrolidone (PVP). Cellular Immunology, 107, 121–129

VAN DEN BOGERT, C., BLAAUW, E.H., DONTJE, B.H.J., HULSTAERT, C.E., HARDONK, M.J. & KROON, A.M. (1986) The effect of doxycycline on polyvinylpyrrolidone-induced granuloma formation in the rat liver. Virchows Arch. (Cell Pathol.), 51, 39–50

VERNIORY, A., DECOODT, P., DUBOIS, R., GASSEE, J.P. & LAMBERT, P.P. (1974) Intraglomerular hemodynamics as estimated from sieving data for polyvinyl-pyrrolidone (PVP) macromolecules in the dog. In Colloque Europeen de physiologie renale, Paris, p. 198

VERNIORY, A., DUBOIS, R., DECOODT, P., GASSEE, J.P. & LAMBERT, P.P. (1973) Measurement of the permeability of biological membranes: application to the glomerular wall. J. Gen. Physiol., 62, 489–507

VILDE, L., CAULET, T., ADNET, J.J., HOPFNER, C. & PLUOT, M. (1968) Considerations histochimiques et ultra-structurales a propos de deux observation de thesaurismose par la polyvinyle pyrrolidone. Arch. Anat. Pathol., 16, 24–35

VOGEL, G. (1977) in Prog. in Lymphol., ed. Magull, R.C. and Whitte, M.H., Plenum (New York), pp. 35–45.

VOGEL, G., & STROCKER, H., (1964) Die Penetration von Polyvinylpyrrolidon durch die Plasma-Lymph-Schranke bei Ratten. Eine Methode zur Beurteilung der Capillarpermeabilität. Pflügers Arch. Ges. Physiol., 279, 187–191

VOGEL, G., & STROCKER, H., (1967) Regionale Unterschiede der Capillarpermeabilität: Untersuchungen über die Penetration von Polyvinylpyrrolidon und endogenen Proteinen aus dem Plasma in die Lymphe von Kaninchen. Pflügers Arch. Ges. Physiol., 294, 119–126

VOGEL, G., ULBRICH, M. & GARTNER, K. (1969) Exchange of plasma albumin (^{131}I-albumin) between extra- and intra-vascular space of the kidney, the lymphatic flow of macromolecules (polyvinylpyrrolidone) in the kidney under conditions of normal and furosemide-inhibited tubular reabsorption. Studies on the function of the renal interstitial tissue and the significance of the tubular reabsorpate for the interstitial tissue. Pflügers Arch., 305, 47–64

VON HENGSTENBERG, J. & SCHUCH, E. (1951) Molekulargewichtsbestimmung von Polyvinylpyrrolidon (PVP) mittels des osmotischen Drucks und der Lichtstreuung ihrer Lösungen. Makromol. Chem., 7, 236–258

WALDMANN, T.A. (1972) Protein-losing enteropathy and kinetic studies of plasma protein metabolism. Semin. Nucl. Med., 2, 251–269

WEIKEL, J.H. & LUSKY, L.M. (1956) Pharmacology of the reticuloendothelial system. 1. Blockade by polyvinylpyrrolidone (PVP) as measured with radio-chromic phosphate in the rabbit. J. Pharmacol. Exp. Therap., 118, 148–152

WESSEL, W., SCHOOG, M. & WINKLER, E. (1971) Polyvinylpyrrolidone (PVP)–its diagnostic, therapeutic and technical application and consequences thereof. Arzneim. Forsch., 21, 1468–1482

WHO (1979) N-vinyl-2-pyrrolidone and polyvinyl pyrrolidone. IARC Monographs on the Evaluation of the Carcinogenic Risk of Chemicals to Humans. Vol. 19, 461–477

WHO (1980) 24th report of the Joint FAO/WHO Expert Committee on Food Additives, Rome, Wld. Hlth. Org. techn. Rep. Ser., No. 653

WHO (1981) 25th report of the Joint FAO/WHO Expert Committee on Food Additives, Geneva, Wld. Hlth. Org. techn. Rep. Ser., No. 669

WHO (1986) 30th report of the Joint FAO/WHO Expert Committee on Food Additives, Rome, Wld. Hlth. Org. techn. Rep. Ser., No. 751

WIDGREN, S. (1965) Thesaurismose ganglionnaire artificielle apres polyvinyl-pyrrolidone. Frankfurt Z. Pathol., 74, 754

WILKINSON, A.W. & STOREY, I.D.E. (1954) Urinary excretion of polyvidone. Lancet, 2, 1269–1271

WOLVEN, A. & LEVENSTEIN, I. (1957) One year feeding study in dogs with PVP. Unpublished report from Leberco Laboratories (USA). Submitted to WHO by GAF.

WUNSCH, P.H. & KIRCHNER, T. (1981) Polyvinylpyrrolidon-granulom. Eine differentialdiagnostisch bedeutsame Weichgewebsläsion? Herbsttagung der Deutschen Gesellschaft fur Pathologie, 12

YAMASAKI, H., ENDO, K. & SAEKI, K. (1969) Plasma histamine and coagulation time of the blood in dogs after administration of different histamine releasers. Acta Med. Okayama, 23, 453–464

YOULTEN, L.J.F. (1969) The permeability to human serum albumin (HSA) and polyvinylpyrrolidone (PVP) of skeletal muscle (rat cremaster) blood vessel walls. J. Physiol., 204, 112P-113P

ZEIGER, E., ANDERSON, B., HAWORTH, S., LAWLOR, T., MORTELMANS, K. & SPECK, W. (1987) Salmonella mutagenicity tests: III. Results from the testing of 255 chemicals. Environ. Mutagen., 9 (Suppl. 9), 1–110

ZELLER, H. & ENGLEHARDT, C. (1977) Testing of Kollidon 30 for mutagenic effects in male mice after a single intraperitoneal application, Dominant Lethal Test. Unpublished report from BASF Gewerbehygiene und Toxikologie. Submitted to WHO by BASF

ZELLER, H. & PEH, J. (1976a) Report on a study on the effects of Kollidon 25, Batch 1229, on the prenatal toxicity with rats. Unpublished report from BASF Gewerbehygeine und Toxikologie. Submitted to WHO by BASF

ZELLER, H. & PEH, J. (1976b) Report on a study on the effects of Kollidon 90, Batch 5, on the prenatal toxicity with rats. Unpublished report from BASF Gewerbehygeine und Toxikologie. Submitted to WHO by BASF

ZENDZIAN, R.P. & TEETERS, W.R. (1970) Acute intravenous administration of PVP in beagle dogs. Unpublished report by Hazleton Laboratories for National Cancer Institute

ZENDZIAN, R.P., TEETERS, W.R. & KWAPIEN, R.P. (1981) Acute intravenous administration of polyvinylpyrrolidone (PVP) in Rhesus monkeys. Unpublished report by Hazelton Laboratories for National Cancer Institute

ZIMECKI, M. & WEBB, D.R. (1978) The influence of molecular weight on immunogenicity and suppressor cells in the immune response to polyvinylpyrrolidone. Clin. Immunol. and Immunopathol., 9, 75–79

Appendix

Appendix 1. Absorption of PVP

A. Rat Studies

Species	Method	M.W.	Observations
Rat (1)	3.5% ^{14}C-PVP orally to 5 rats (6–10g/kg). Urine, feces, and expired CO_2 and carcass radioactivity measured.	40,000 (K-30)	Large dose of PVP caused diarrhea making collection of total fecal RA difficult. 1% collected in urine in first 24 hours, 0.25% in CO_2 and 0.5% in carcass. The rest assumed to be excreted in feces.
Rat (2)	10% ^{14}C-PVP orally to a total of 4 rats (2–2.5g/kg) Urine, feces, CO_2 and tissue radioactivity measured.	K-90	Urine collected with funnel taped to penis in 2 rats ($<0.35\%$ in 13 hours). 0.04% of RA in CO_2 over first six hours then undetectable. Urine and feces accounted for 97%. Liver and G.I.T. accounted for 0.012% and 0.003%.
Rat (3)	5.0 mg/kg of ^{14}C-PVP orally to 4 groups of 5 rats. A group killed at each of 6, 12, 24 or 48 hrs. Urine, feces and tissues collected and RA measured.	K-30	Cumulative fecal recovery at 42 hrs was 90.8% rising to 98.4% at 48 hours. Only background counts detected in all tissues. 0.004% of administered dose found in urine during first 6 hours. Minute amounts detected in blood stream reaching peak by 2 hours. Absorption complete by 6 hours. (Plasma half life = 1.5 hrs). Dialysis studies show 4% of PVP has M.W. $<$ 3,500. Probably accounts for absorbed RA.
Rat (4)	50 mg/kg ^{14}C-PVP given intraduodenally to anaesthetised rats. Urine and bile excretion measured.	1,700 (K-12) 40,000 (K-25)	Relative proportions of K-12 and K-25 absorbed compared from amounts excreted in urine and bile. 1.3% of K-25 excreted in urine and bile. 7.55% of K-12 excreted in urine and bile.
Weanling rat (5)	^{125}I-PVP (5 mg/rat) given by stomach tube. 4 hours later intestine washed and retention of RA in intestine measured.	160,000 (K-60)	Weanling rats less than 18 days old retain up to 50% of RA in intestine probably by pinocytosis. The process is saturable with uptake being proportional to dose up to 1 mg/rat. Between 18–20 days sharp decline in PVP uptake coinciding with change in histological appearance of intestine (called closure). Adult rats take up $<$ 1%.

1. Shelanski (1953)
2. Shelanski (1960)
3. Digenis et al (1987)
4. Morgenthaler (1977)
5. Clarke & Hardy (1969a, 1969b, 1971)

Appendix 1. Absorption of PVP

B. Other Studies

Species	Method	M.W.	Observations
Rabbit (1)	^{131}I-PVP infused into intestine of anesthetised rabbits. Blood samples collected and perfusate and plasma fractionated on G200 Sephadex to identify different MW PVP fractions	MW range 8000–80,000	Inverse linear log/log relationship between permeability and MW. Plasma/perfusate ratio for PVP with MW of 10,000 about 0.09, for MW 22,000 is 0.025 and MW 50,000 about 0.01. It has been calculated that PVP's with MWs of 8,000, 33,000, and 80,000 would cross the intestine at 0.67%, 0.39% and 0.067% of the rate of urea.
Rabbit (2)	A. 70g of PVP as dietary adjunct over 1 month. B. Perfusion of small intestine in anaesthetised rabbits with 7% PVP.	40,000	A. Up to 30 mg of PVP recovered in liver corrected for recovery rate. B. 14g administered – maximal absorption after 10 mins – by 30 mins blood level 10% of peak.
Human (3)	^{14}C-PVP given orally to patients receiving 5-fluorouracil to assess changes in GIT permeability	Av.MW 11,000	Baseline permeability before and 5 weeks after therapy showed absorption of 0.013–0.048% of administered PVP. This increased 2–20 fold as a result of intestinal damage caused by 5-fluorouracil. PVP is adsorbed to stools. PVP in urine has lower MW range than that administered.
Sheep (4) (ewe)	^{131}I-PVP administered into abomasum and duodenum.	33,000	When given into abomasum 60% of RA recovered in feces in 20 hours. When given into duodenum 71% of RA recovered in feces in 20 hours. The missing radioactivity was assumed to be absorbed, but no figures given.
Calf & pig (5)	^{131}I-PVP infused into the duodenum of anaesthetised animals.	160,000	PVP detectable in lymph following absorption from G.I.T. in newborn animals.

1. Loehry (1970)
2. Haranaka (1971)
3. Siber et al (1980)
4. Fell et al (1969)
5. Hardy (1965, 1968)

Appendix 2. Renal Elimination of PVP

A. Animal Studies – Rat

Species	Method	M.W.	Observations
Rat (1)	Studied movement of PVP through glomerular and post-glomerular capillaries	11,500 110,000 650,000	Glomerular capillaries just allow molecules of MW 25,000 to pass. Larger molecules exit through post-glomerular capillaries. Some reabsorption via same capillaries or lymphatics can occur.
Rat (2)	Gave ^{14}C-PVP i.v. and measured appearance in urine in anaesthetised rats.	K-12 (1,700) K-17 (6,300)	K-12 cleared at 100% G.F.R. K-17 cleared at 50% of GFR. Clearance not affected by plasma concentration over range 5–10,000 nmol and not affected by diuresis. Suggest entirely glomerular filtration. Active transport and secretion excluded. Endocytic uptake of PVP by lysosomal system in tubule (0.3–0.4% of dose).
Rat (3)	50–200 mg/kg given i.v. and urinary excretion monitored over 3–22 days	K-14 K-18	93% of K-14 and K-18 excreted in urine over 72 hrs (mostly first 24 hrs) when given at 50 mg/kg. At 200 mg/kg 92% of K-14 and 86% of K-18 collected over 22 days.
Rat (4)	i.v. PVP	K-12 K-17	98% of K-12 and 94% of K-17 eliminated over 10–27 days. More than 90% collected in first 24 hours.
Rat (5)	oral ^{14}C-PVP	40,000	Less than 0.04% of RA found in urine within 6 hrs. After that PVP not detectable.

1. Gartner et al (1968)
2. Schiller and Taugner (1980)
3. Hespe (1977)
4. Indest and Brode (1977) Indest (1978)
5. Digenis et al (1987)

Appendix 2.   Renal Elimination of PVP   B. Animal Studies—Other Species

Species	Method	M.W.	Observations
Rat/Dog/Rabbit/ Man (1)	Chemical estimation of PVP in urine following i.v. administration to a number of different species.	Various MW 11,600– 500,000	Review of world literature showed that rate and extent of renal elimination inversely related to MW. Half-life elimination for PVP MW 40,000 ranged from 12-72 hours.
Dog (2)	^{131}I-PVP given i.v. to anaesthetised dogs and cumulative urinary excretion measured.	K-32 K-35 K-50	90% K-32, 65% K-35 and 15% of K-32 appeared in urine within 72 hours. Most within 24 hours.
Dog (3)	Plotted sieving coefficient of glomerulus against radius of PVP molecule after i.v. injection.	Various	Distribution of glomerular pores 2.38-6.31 nm (mean 3.9 nm) suggests PVP MW < 8000 will clear glomerulus at 100% GFR. Pore size also suggests that PVP MW 39,000 should be cleared, but molecules above 104,000 will be excluded.
Dog (4)	PVP i.v. to 2 female dogs. Urine collected for 36 hours and fractionated to determine MW.	38,000	Small molecules appeared in the urine before large. Largest molecules retrieved from urine had MW 87,000-104,000.
Rabbit (5)	Fractionated ^{14}C-PVP equivalent to K 17.8 given intravenously.	MW range 20-50,000	72% collected in urine in 24 hrs, 89% in 48 hrs and 94% within 14 days.

1. Wessel et al (1971)
2. Ravin et al (1952)
3. Lambert et al (1970, 1971, 1972)
4. Stahl and Frauenfelder (1972)
5. Siber et al (1980)

Appendix 2.   Renal Elimination of PVP   **C. Human Studies—Healthy Subjects**

Species	Method	M.W.	Observations
Normal subjects (1)	i.v. infusion over 40–45 mins of PVP with large MW distribution.	K-27 K-33 K-42	Viscosity measurements of collected PVP during first 6 hours shows lower K-range than injected material. K-value rises over 48-96 hr period.
Normal subjects (2)	^{125}I-PVP i.v. then serial 1 hr plasma and urine samples. Fractionated to determine MW.	38,000	Increasing urine/plasma concentration ratio with decreasing molecular size of polymer. Relationship continued up to limiting value approximating to creatinine urine/plasma concentration ratio.
Normal subjects (3)	^{125}I-PVP injected i.v.	Various	By fractionation of PVP entering urine it was shown that healthy people can excrete PVP with molecular radius as high as 6 nm (equivalent to MW of 94,000).
Normal subjects (4)	^{125}I-PVP injected i.v.	38,000	Fractionation of PVP in urine showed glomerulus to be highly permeable to molecules with MW <30,000, and relatively impermeable to molecules > 70,000.
Normal subjects (5)	Sieving coefficient of glomerulus plotted against radius of PVP appearing in urine after i.v. injection.	Various	Distribution of pore-size 1.8-6.7 nm (Mean 3.4 nm). The mean pore size would allow PVP with Av MW of 31,000 to be cleared and maximal pore size would allow molecules up to 116,000 to pass through to some extent.

1. Ravin et al (1952)
2. Hulme and Hardwicke (1968), Hardwicke (1972)
3. Hulme (1975)
4. Blainey (1968)
5. Lambert et al (1970)

Appendix 2. Renal Elimination of PVP D. Human Studies – Patients

Species	Method	M.W.	Observations
20 patients in shock requiring plasma expander(1)	i.v. 3.5% PVP given in volumes up to 1,500 ml.	Plasmosan (29,800–56,000)	PVP analysed chemically in urine until only trace amounts detected (up to 260 hrs). > 70% excreted by 5 patients, 50-70% in 9 patients, 40-50% in 5 patients and <30% in 1 patient. No obvious signs of toxicity or accumulation in major organs.
Patients with metastatic carcinoma(2)	^{14}C-PVP given orally. Urine, fecal and plasma levels of RA monitored	11,000	Part of study to measure gastro-intestinal absorption of PVP during treatment with 5-fluorouracil. Peak urinary excretion occurred during first day. Total urinary excretion of oral dose was 0.013-0.04%.
Normal and renal transplant patients(3)	^{125}I-PVP given i.v. Urine and plasma collected and fractionated.	38,000	Results suggest that maximum molecular size that could cross healthy glomerulus would have radius of <4.0 nm.
Patients requiring plasma expander(4)	i.v. infusion of 540ml of PVP. Urine collected over 14 days.	35,000	60% of PVP excreted in 24 hours, 80% within 14 days. Multicompartment system suggested to explain pharmacokinetics.
Terminally ill cancer patients(5)	Iodine-labelled PVP injected i.v. (19g)	K-28/32 MW 40,000	Elimination of urine monitored and tissue levels estimated at autopsy 2-8 weeks later. Within 6 hrs 33% excreted, with 24 hrs 66% excreted. Unexcreted PVP in greatest amounts in kidney, liver, spleen and lymph nodes. No adverse histology noted.
14 child patients(6)	45ml of 6% PVP injected i.v.	10,000	95% of PVP excreted in urine in 6 hours.

1. Wilkinson and Storey (1954)
2. Siber et al (1980)
3. Robson et al (1973)
4. Campbell et al (1954)
5. Loeffler and Scudder (1953)
6. Brautigam and Gleiss (1956)

Appendix 3. Distribution of PVP

A. Pharmacokinetic Studies

Species	Method	M.W.	Observations
Rats, dogs rabbits, humans (1)	Radiolabelled PVP injected i.v.	Various	Rate of disappearance related to MW. Initial rapid fall in blood level due to urinary excretion, followed by slower decay due to passage into lymph and uptake by tissues. MW < 25,000 readily cleared by glomerulus. MW > 110,000 retained by RES "possibly for years."
Rats (2) Rabbits (3)	PVP infused i.v.	11,000 38,000 90,000	Passage from plasma to lymph dependent on MW. Lymph/plasma ratio of PVP in rabbit is different in each tissue (hind legs = 0.3, thoracic duct = 0.35, kidney = 0.7, liver = 0.8). Protein lymph/plasma ratio remain about the same.
Sheep (4)	PVP infused i.v.	K-11 K-12	Lymph/plasma ratio in lungs depends on molecular size.
Rat (5)	Superfused rat cremaster muscle used to determine blood vessel permeability.	K-30	Passage of PVP through blood vessels depends on molecular size. Critical size is between 2.5–3.1 nm (MW 16,000–25,000). Means that volume of distribution is dependent on MW.
Rabbit (6)	Filtration coefficients of PVP measured in various tissues after i.v. injection	Range from 11,500 to 650,000	Only traces of PVP with MW < 25,000 found in lymph. MW > 25,000 found in liver and kidney. Significant filtration shown in liver with PVP of MW 650,000.
Rabbit (7)	Compared plasma and whole body RA decay curves to estimate uptake by RES.	33,000	By allowing time for renal excretion of low MW PVP showed significant uptake of higher MW by RES. Plasma half-life = 3.2 hours. Whole body half-life = 18.0 hours.

1. Ravin et al (1952)
2. Vogel and Strocker (1964)
3. Vogel and Strocker (1967)
4. Boyd et al (1969)
5. Youlten (1969)
6. Vogel (1977)
7. Regoeczi (1976)

Appendix 3. Distribution of PVP

B. Tissue Uptake Studies

Species	Method	M.W.	Observations
Rabbit & Rats (1)	^{14}C-PVP injected i.v. and tissues removed and counted for RA	K-33	Animals killed at various times after dosing. Largest amounts of PVP found in skeletal muscle (5–17%) and skin (5–10%). Highest relative concentrations found in liver, spleen, bone marrow and lymph nodes. Retention of PVP by tissues increased with increasing MW (due to slow excretion).
Rabbit (2)	0.8g/kg ^{14}C-PVP injected i.v.	K-28/32	Highest concentrations of PVP found in kidneys, lungs, liver, spleen and lymph nodes.
Rabbit (3)	Organ radioactivity measured 10 days after i.v. injection of ^{131}I-PVP	36,000 with low MW material removed.	Of radioactivity still present in carcass 17.5% in blood 21% in skin, 15% in liver, 10% in bone marrow and smaller amounts in other tissues.
Rabbit (4)	^{14}C-PVP injected i.v.	Fractionated 11,000 (K-17.8)	Selective uptake by liver and spleen, immediately after injection and return to background level within 12 weeks i.e. RES uptake for MW 11,000 is reversible.
Rat (5)	^{3}H-PVP (100mg/kg) injected i.v.	30,000	48 hrs after dosing liver contained 35% of PVP, muscles 11%, spleen 6%, kidneys 1.8%, lungs 0.7% and blood 1.8%. With larger doses disproportionately large amounts taken up by spleen.
Rat (6)	^{14}C-PVP i.v.	K-14 K-18	Lower tissue uptake of K-14 compared with K-18 in all tissues except kidney and gastrointestinal tract.
Rat (7)	^{14}C-PVP injected i.m. Muscle and popliteal lymph node RA measured.	K-12 K-17 K-30	Most radioactivity excreted within 3–4 days. Muscle radioactivity fell steadily over 18 day period. Initial uptake of RA into lymph nodes dependent on MW. No consistent change in RA of nodes over 78 day period.

1. Ravin et al (1952)
2. Loeffler and Scudder (1953)
3. Regoeczi (1976)
4. Siber et al (1980)
5. Marek et al (1969)
6. Hespe et al (1977)
7. Cameron and Dunsire (1983a, 1983b).

Appendix 3.   Distribution of PVP

C. Histological Studies

Species	Method	M.W.	Observations
Rabbit (1)	Repeat i.v. injection of PVP over months until total dose about 30g	—	Histological picture characteristic of lipid storage disease. No change in picture over observation period of 41 weeks.
Mouse (2)	0.5 ml of 3.5% or 20% PVP injected i.v. on number of occasions over 6 days.	20,000 40,000 125,000	"Foam cells" developed in liver only after multiple injections and when more than 0.1g/kg PVP injected. For PVP with MW 40,000 this did not occur until volume of PVP injected exceed blood volume. Number of foam cells related to MW in ratio 6:2:1 for high medium and low MW PVP.
Rat (3)	Infused 75 ml of 3.5% PVP solution.	40,000	Foamy appearance noted in endothelial cells, tubular epithelial and glomerular epithelial cells of kidney, in Kupffer cells in liver and macrophages in adrenal cortex.
Rat (4)	PVP injected i.v. over 3 days at 0.5-2.5g/kg	34,000 270,000	Fluorescence microscopy showed intracellular material (assumed to be PVP) in the liver, spleen, lung and kidney through 1 year of observation.
Rabbit (5)	PVP given i.v. at intervals for 1 month	12,000 28,000 220,000	With K-20 isolated accumulation of PVP in liver and spleen. With higher MW PVP accumulation in these organs but also heart, lungs and adrenals.
Rat (6)	PVP (0.2-2.0 mg) injected i.m. up to 5 times	K-12 K-17 K-30 K-90	Acute muscle damage at 3 days in all groups including control. Granulomatous reaction in some animals, but not thought to be peculiar to PVP. No lymph node changes in any groups.

1. Ammon and Muller (1949)
2. Frommer (1956)
3. Upham et al (1956)

4. Mohn (1960)
5. Oettel and von Schilling (1967)
6. Merkle et al (1983)

Appendix 3.   Distribution of PVP   D. Intravenous Administration to Humans

Species	Method	M.W.	Observations
Patients(1)	6 ml of 10% radiolabelled PVP	40,000	No histological evidence of PVP accumulation, but RA found in kidneys, lungs, liver, spleen and lymph nodes.
25 patients (2)	Single infusion of 1 litre of PVP (3.5–4.5%)	40,000	Serial liver biopsies before and up to 13 months after infusion. Basophilic globular deposits 50 μ in diameter, rare before 3 months, but common after 6 months.
23 patients (3)	Clinical and laboratory follow-up of 23 of above patients.	40,000	Slight or moderate elevation of RBC and haemoglobin level. Temporary elevation of cephalin flocculation in 11 patients. No change in thymol turbidity, total lipids or serum bilirubin.
19 patients (4)	–	–	Similar storage changes to PVP noted in kidney to those seen after dextran, sucrose and gelatin.
300 patients (5)	Up to 28 litres of PVP over 14 day period.	40,000	Some individuals followed up for up to 3 years. Persistent storage noted in spleen, bone marrow, kidney and liver.
144 patients (6)	3.5% PVP as plasma expander.	40,000	Minor complications noted in 49 patients, 17 minor complications below 69g, 43 between 70–99g, 65 between 100–139g, 92 above 140g.
34 autopsy cases (7)	Morphological study in individuals who died from various conditions after PVP exposure	24,800 12,600	Apparent PVP storage characterised by foam cells and/or vacuolar amorphous deposits. Effect dose related. Histological changes in kidney. PVP also localized in inflammatory sites, wounds and neoplastic lesions.

1. Loeffler and Scudder (1953)
2. Gall et al (1953)
3. Altemeier et al (1954)
4. Traenckner (1954a)
5. Traenckner (1954b)
6. Honda et al (1966)
7. Kojima et al (1967)

Appendix 3. Distribution of PVP

E. Mechanism of Uptake

Species	Method	M.W.	Observations
Pregnant rats (1)	PVP injected i.p. at different times during pregnancy.	40,000	If PVP injected on 16th day of gestation, epithelial cells of embryonic yolk sac highly vacuolated 36 hrs later.
Rat peritoneal macrophages (2)	Macrophages incubated in vitro with ^{125}I-PVP	30,000–40,000	Constant linear uptake of PVP over 10 hour period. 10–50% loss of label if incubated with 'cold' PVP.
Weanling rats (3)	^{125}I-PVP given by stomach tube	K-60	PVP taken up into intestinal epithelium by pinocytosis until rat reaches 20 days old. In adult uptake < 1%.
Rat (4)	^{14}C-PVP injected i.v.	K-12	Uptake into kidney shown by autoradiographs. ^{14}C-PVP found in large lysosome-like structures after 6 hrs.
Rabbit (5)	^{131}I-PVP injected i.v.	33,000	Removal of RA from plasma enhanced by injection of materials known to stimulate RES phagocytic activity e.g. fibrinogen, sterile inflammation, immune reactions.

1. Roberts et al (1976)
2. Pratten et al (1977)
3. Clarke and Hardy (1969a, 1969b)
4. Schiller and Taugner (1980)
5. Regoeczi (1976)

Appendix 4.   Acute Toxicity of PVP

Species	Route	M.W.	Observations
Rat (1)	Oral	40,000	LD 50 = 100g/kg
Rat (2)	Oral	10,000–30,000	
Guinea pig (1)	Oral	40,000	LD 50 = 40g/kg
Mouse (2)	Oral	–	LD 50 = 100g/kg
Mouse (3)	i.p.	–	LD 50 = 40g/kg
			LD 50 = 12–15g/kg
Rabbit (5 male and 5 female/grp) (4)	orally by gavage	50,000	300, 900 or 2700 mg/kg of PVP. Controls received water or hydroxypropyl cellulose (HPC). Slight inhibition of body weight gain in high dose group receiving PVP or HPC. No treatment related changes in plasma enzymes or protein, in liver function or in liver histology.
Beagle dog (1/sex/grp) (5)	i.v.	40,000	0, 1, 3 or 10g/kg in 50ml/kg. Shock reaciton at 10g/kg in which dogs showed tremors and/or sub-convulsive movements, defaecation, salivation, depression and ptosis (thought to be due to hypertonic nature of soln). During 28 days of observation there were sporadic changes in hematology and transaminases, but all returned to normal within 48 hrs. No histopathological or gross changes reported.
Rhesus monkey (6)	i.v.	15,000	10g/kg as 20% solution killed animal after only 53% injected. Thought to be due to viscous hypertonic nature of solution. Animals survived 5g/kg as 10% solution with some minor transient changes in blood chemistry. Animals killed 28 days later. No compound-related histological changes.

1. Shelanski et al (1954)
2. Scheffner (1955)
3. Angervall & Berntsson (1961)
4. Neumann et al (1979)
5. Zendzian & Teeters (1970)
6. Zendzian et al (1981)

Appendix 5. Subchronic Toxicity of PVP

A. Rat Studies

Species	Route	Duration	M.W.	Observations
Rat (1)	10g/kg in diet	9 wks	10,000	6 animals died from non-drug associated causes at intervals throughout study. 4 rats sacrificed after 9 wks. Gross autopsy showed minor renal irritation in 1 rat. Intestinal progress in PVP fed animals was slower than in control.
Rat (1) (4/grp)	3% PVP solution	24 wks	11.500	No difference in weight gain between control and test groups.
Rat (1) (11 rats)	10g/kg? in diet	9.5 wks	30,000	Gross pathology of survivors showed no significant toxicological changes. No effect on passage of barium meal.
Rat (2) (11 rats)	10g/kg 48 × gavage	2 mths	K-30	The 7 surviving animals showed excipient bronchial pneumonia and severe desquamation of cells in gastric mucosa.
Sherman Wistar rat (25/sex/grp) (3)	0, 2%, 5% or 10% PVP in diet	90 days	K-90	No significant differences in weight gain curves. No consistent changes seen on gross or microscopic examination of tissues. Special staining techniques for PVP were negative.
Sprague (4) Dawley rat (10/sex/grp)	0, 2.5% or 5% PVP in diet	28 days	K-90	No toxic effects or pathological-histological findings related to PVP administration.

1. WHO (1980) and unpublished
2. Scheffner (1955)
3. Shelanski (1959a)
4. Kirsch et al. (1972)

Appendix 5. Subchronic Toxicity of PVP   B. Cat and Dog Studies

Species	Route	Duration	M.W.	Observations
Cat (1) (7 cats)	5g/kg by gavage	5 days/wk for 7 wks	10,000	All animals survived. No adverse effects on amount of urine excreted. No hematological findings.
Cat (1)	5g/kg	5 days/wk for 5.5 wks	30,000	Two animals died from unconnected causes. No significant gross pathology or histology in remainder attributable to PVP. Urine and blood normal.
Cat (2)	—	—	<40,000	High doses caused diarrhea–minimum effective dose 0.5g/kg.
Dog (2) (2 dogs)	5g/kg orally	1.5 wks 2 wks	220,000 1,500,000	No abnormal findings.
Dog (1) (4 dogs)	Fed 5g/kg	5 days/wk for 5 wks	10,000	No deleterious effects on amount and composition of urine. Blood normal 3 hours after intubation. Soft feces appeared temporarily at levels above 1g/kg.
Dog (1) (2 dogs)	5g/kg body weight	9 wks	20,000	No toxic effects noted. Blood and urine values remained normal.
Beagle dog (3) (2/sex/grp, 16 total)	Diet alone 2% PVP 5% PVP 10% PVP	90 days	K-90	High dose group lost weight significantly compared with control animals (which gained weight). No consistent treatment-related pathology. Special PVP staining of mesenteric lymph nodes was positive in 3/4 in 10% grp, 1/4 in 5% grp, 1/4 in 2% grp, 1/4 in controls.
Beagle dog (4) (4/sex/grp, 40 total)	0, 2.5%, 5% or 10% PVP or 5% cellulose in diet	28 days	360,000 (K-90)	No toxic effects or pathological changes which could be related to administered substance. Except slightly increased spleen weight in female animals of 10% PVP group.

1. WHO (1980) & unpublished
2. Scheffner (1956)
3. Shelanski (1956b)
4. Kirsch et al (1975)

Appendix 6. Chronic Toxicity of PVP

A. Rat Studies

Species	Route	Duration	M.W.	Observations
Sherman Wistar rat (1) 50/sex/grp	0, 1 or 10% PVP in diet	2 yrs	37,900 (K-30)	Part of carcinogenicity test. Monthly weight records showed 10% PVP group within 10% of control curve. More frequent and more fluid stools in 10% PVP group. Haematology within normal range and no result attributable to diet. 3 month urine analysis (from 5 rats) showed normal records during first 15 months. At 18 months albumen present in 10% PVP group. At 21st month appeared in all groups. 10 rats/grp autopsied at 24 months. Special attention given to lymphatics on these and on any animals that died. No treatment-related changes in histopathology or in gross pathology.
Sprague Dawley rat (2) 50/sex/grp	0, 5, or 10% PVP or 5% cellulose in diet	2 yrs	30,000 (K-25)	Part of carcinogenicity study. Body weight measurements showed 10% PVP did not affect growth in males or females. No observable toxicity in any groups. Food intake, blood and urine picture all within normal limits (except SGPT low in 10% PVP group at 4th week only). No associated pathology. No evidence of PVP storage in mucous membranes of duodenum or intestine, or in the mesenteric lymph nodes.
Sprague Dawley rat (3)	1, 2.5 or 5% PVP or 5% cellulose in diet	129 wks in males 138 wks in females	K-90	Part of carcinogenicity test. 150 rats/group on treatment with 250 animals as controls. 5 animals/sex/group killed at 26, 52 and 104 weeks. No toxicity reported and no signs of abnormal storage of PVP in liver, kidney, heart or lymph nodes.

1. Shelanski (1957)
2. BASF (1978)
3. BASF (1980a)

Appendix 6.   Chronic Toxicity of PVP

B. Dog Studies

Species	Route	Duration	M.W.	Observations
Dog (1) (32 in total)	In diet 5% or more	1 yr	37,900 (K-30)	Two studies reported together. No observable toxic effects. No malignant tumours but slight evidence of PVP absorption from GI tract. Special staining technique for PVP gave positive staining in mesenteric lymph nodes of all animals (including controls) but negative findings in liver, intestine, and spleen of all animals. PVP detected in blood of test animals but of doubtful significance since also found at similar levels in controls.
Beagle dog (2) 2/sex/grp	In diet 10% PVP + 0% SF 5% PVP + 5% SF 2% PVP + 8% SF 10% Solka-Floc (SF) (SF = cellulose)	2 yrs	37,900 (K-30)	All animals remained healthy with no adverse effects on body weight gain, blood or urine analysis. At autopsy no gross or histological pathology in a wide variety of organs: except swollen reticuloendothelial cells in 10% PVP group. This was also reported in lower dose groups but less consistently and to a lesser degree.

1. Shelanski (1958) and Wolven & Levenstein (1957) reported by Burnette (1962)
2. Princiotto et al (1954)

Appendix 7. Teratogenicity of PVP

Species	Procedure	M.W.	Observations
Sprague Dawley rat (1) (25/grp)	10% PVP in diet for 20 days after evidence of pregnancy. Controls received diet alone. Caesarean section on Day 20	25,000 (K-25)	Body weight of rats receiving PVP increased less than controls but fetuses in this group slightly heavier. No gross pathology in mothers. Conception rate unaffected (100% PVP, 96.3% controls). No dead fetuses in either group. Skeletal malformation rate in controls 3.34% and in PVP group 0.67%. The number of implantations and resorptions were similar.
Sprague Dawley Rat (2) (30/grp)	10% PVP in diet for 20 days after evidence of pregnancy. Controls received diet alone. Caesarean section on Day 20	K-90	Body weight of rats receiving PVP increased less than controls but fetuses in this group slightly heavier. No pathology in mothers attributable to PVP. Similar number of implantations and resorptions in each group. Number of fetuses similar in controls (329) and PVP group (352). No dead fetuses and similar distribution of malformations.
Rabbit (3) (11-12/grp)	Animals artificially inseminated and 50, 250 or 1250 mg/kg PVP given i.v. daily from Day 6-18. Untreated and saline controls. Caesarian section Day 28.	10,000 (K-12)	No pathology in mothers and no consistent change in body weight gain. Conception rate (%) 90.9 (saline)—100 (250mg/kg) Corpora lutea 7.82 (untreated)—9.64 (50 mg/kg) Implantations 5.83 (250 mg/kg)—7.11 (saline) Viable fetuses 4.92 (250mg/kg)—6.67 (saline) Live implantations (%) 90.6 (250 mg/kg)—93.4 (saline) There were no significant variations in fetal weight (except high dose PVP which were heavier). Only one malformation seen and that was in the saline group.
Rabbit (4)	550 μg PVP injected into yolk sac of 9 day embryo	11,500	No increase in number of resorptions or malformations compared with saline injected embryos.

1. Zeller & Peh (1976a)
2. Zeller & Peh (1976b)
3. Hofman & Peh (1977)
4. Claussen & Breuer (1975)

Appendix 8. Mutagenicity of PVP

A. In Vitro Studies

Species	Procedure	M.W.	Observations
Ames test (1)	3.5% PVP tested on Salmonella typhimurium mutant without tissue activation.	37,900	No samples caused reversion of histidine auxotrophic mutants to synthesise histidine freely and are therefore regarded as not mutagenic.
Ames test (2)	PVP in hair spray resin tested on Salmonella typhimurium (Strains Ta 1537, Ta 1538 and Ta 98) with liver microsome activation.	37,900	At levels up to 5mg/plate none of the samples showed an increase in revertent colonies over background control plates.
Mouse lymphoma (3)	PVP 0.1%–10% and PVP-I 0.01%–1% compared with ethylmethane sulphonate (known mutagen) on L 5178Y strain mouse lymphoma cells	K-30	No PVP treatment groups had a biologically significant effect on cell viability or mutational frequency. PVP-I had no significant effect except at 1% which was cytotoxic.
Mouse cell (Balb/c 3T3) transformation (3)	PVP 0.5%–10% and PVP-I 0.01%–1% compared with ethylmethane sulphonate and MNNG for ability to induce transformation of non-malignant (non-contact-inhibited) cells into malignant cells.	K-30	No PVP samples had significant effect on cell viability or mutational frequency compared with controls. PVP-I at concentrations which were not cytotoxic had no effect on transformation.

1. Bruce (1977)
2. Clairol Laboratories (1978)
3. Kessler et al (1980)

Appendix 8.   Mutagenicity of PVP

B. In Vivo Studies

Species	Procedure	M.W.	Observations
Dominant Lethal Test (1)	Single i.p. inj. into male mice of 3.16g/kg in a volume of 10 ml/kg. Untreated & vehicle controls.	K-30	No animals of either sex showed signs of toxicity. Rate of conception, average number of implantations, percentage live fetuses, and mutagenic index all similar to controls.
Bone marrow chromosomal aberration (2)	Groups of 5 Chinese hamsters/sex given PVP 3.16g/kg i.p. Groups of 3 injected with 10 ml/kg distilled water as controls. At 4, 22 & 46 hrs animals injected i.p. with colcemide 3.1 mg/kg. 2 hrs later femoral bone marrow collected, stained and 100 metaphases/animal examined for chromosomal aberration.	Kollidon 30	Under experimental conditions no clinically noticeable symptoms and pathological changes of internal organs in male or female hamsters caused by PVP. No chromosomal aberrations in bone marrow cells of treated animals.

1. Zeller & Engelhardt (1977)
2. BASF (1980b)

Appendix 9. Carcinogenicity of PVP

A. Hueper Studies

Species	Route	Duration	M.W.	Observations

C 57 mouse (50/grp) Bethesda black rat (1) (15-30/grp)	s.c. & i.p. implantation of powder 200 mg/mouse, 500 mg/rat i.v. inj. of 7% soln. in rat, 2.5 ml once/week for 8 weeks	2 yrs.	4 samples in range 20,000–300,000	The following overall incidence of tumours was reported with PVPs of differing MWs:
				
				    RES sarcomas Carcinomas
				    Rats Mice Rats Mice
				s.c. 23/120 4/200 2/120 0/200
				i.p. 26/120 4/200 4/120 0/200
				i.v. 11/60 — 4/60 —
				
				The incidence of RES sarcomas in "control" rats was 12/266 (5%) but not all these animals were untreated. The incidence in control mice was 0.4%.

Rat (20-30/grp) Mouse (30-50/grp) Rabbit (2) (2-6/grp)	As powder and 6% soln. 200 mg/mouse (i.p. & s.c.) 500 mg/rat (i.p., s.c. & i.v.) 1,700-6250 mg/rabbit i.v.	2 yrs in rats and mice. 4 yrs in rabbits	10 samples in range 10,000–300,000	Storage reaction characterized by bluish-stained swollen cells in various tissues in all species regardless of MW of PVP, but dependent on amount given. It was most marked in rabbits after i.v. injection. Many tissues storing PVP displayed hyperplastic and preneoplastic reactions. Kupffer cells described as "foamy" with difficulty in determining benign or malignant nature of reaction. A total of 122 tumours were recorded in the 1085 PVP-treated animals. Of these 80% were malignant. Highest incidence in lymph nodes (48, of which 37 were reticulum cell sarcomas), uterus (23, or which 17 were reticulum cell sarcomas) and liver (23, or which 17 were reticulum cell sarcomas).
NIH black rat (20-35/grp) Rabbit (3) (2-6/grp)	2.5 ml of 20 to 25% soln. i.p. in rats for 6-10 wks. 50ml of 20% soln i.p. at 2 wk intervals in rabbits for 6-10 wks.	Maximum survival 24 mnths in rats 28 mnths in rabbits	10,000 (K-17) 18,000 (K-25) 50,000 (GAF) 50,000 (BASF)	A storage reaction (bluish staining on histological section) in tissues of all animals receiving PVP except the rabbits injected with K-17 (removed more readily through kidney). The tumour incidence was: PVP RES Sarcoma Carcinoma Type Rat Rabbit Rat Rabbit K-17 3/35 0/6 2/35 0/6 K-25 1/35 0/6 2/35 0/6 GAF 2/20 — 0/20 — BASF 0/30 — 0/30 — Control 2/30 0/2 3/30 0/2

1. Hueper (1957) 2. Hueper (1959) 3. Hueper (1961)

Appendix 9. Carcinogenicity of PVP

B. Rat Studies—Oral Route

Species	Route	Duration	M.W.	Observations
Sherman Wistar rats (1) 50/sex/grp	0, 1 or 10% PVP in diet	24 months	37,900 (K-30)	No treatment-related toxicity or changes in gross or histopathology (see Table 6A). No evidence of carcinogenicity.
Sprague Dawley 50/sex/grp (2)	0, 5 or 10% PVP or 5% cellulose in diet	24 months	30,000 (K-25)	No treatment-related toxicity during study, no evidence of PVP storage, and no PVP-associated pathology (see Table 6A). In all control and test groups incidence of benign and malignant tumours were within normal limits.
Sprague Dawley rat (3)	1, 2.5 or 5% PVP or 5% cellulose in diet	129 weeks in males 138 weeks in females	K-90	150 rats/group on treatments, with 250 animals as controls. No treatment related toxicity (see Table 6A). Experiments continued until 70% mortality recorded in control group. Number of tumours, site, time of onset and number of tumour-bearing animals similar between groups. In fact in high dose group tumour rates lower than control (Male 53% cf 67%: female 76% cf 84%).

1. Shelanski (1957)
2. BASF (1978)
3. BASF (1980b)

Appendix 9.   Carcinogenicity of PVP

C. Rat Studies—Other Routes

Species	Route	Duration	M.W.	Observations
Osbourne-Mendel (20) and Bethesda-black (10) rats (1)	6% solution 1 ml s.c. at weekly intervals	73 wks	Not stated	Tumors (fibrosarcomas of moderate histological malignancy) found at injection site only. No metastases. No sex difference. 43% incidence in PVP group, 43% in carboxymethylcellulose and 17% in Tween 60 groups (controls).
Hybrid rat (2)	200mg/day i.v.	32 mths	Not stated	40 survivors at 1 year. There were tumours attributed to PVP. Only 1 spontaneous uterine carcinoma after 12 months reported.
Wistar rats (3)	2380 rats given i.p. injections of variety of water soluble polymers	Up to 18 months	—	335 animals treated with PVP. Of all animals treated there were 7 benign and no malignant tumours. Author concluded that no high risk of polymer-induced cancer.
Bethesda black rat (4-72/grp) (4)	3 i.p. injections of 25% soln. at monthly intervals (2 ml/100g) Untreated and dist. water controls.	Lifetime	K-17 K-25	No significant difference in incidence of carcinomas and benign tumours between group. In K-25 group 3 sarcomas (males), in vehicle controls one sarcoma and in K-17 grp and untreated control none.
Sprague Dawley rats (female) (12-25/grp) (5)	Single s.c. implantation of 0.5 g of PVP powder (25 rats) Untreated controls (12 rats)	Lifetime	K-30	Average survival time of treated group was 20 months. No sarcomas at injection site. Number of tumours in treated group not significantly different from control.

1. Lusky and Nelson (1957)
2. Danneberg quoted by Shubik and Hartwell (1957)
3. Lindner (1960)
4. BASF (1960)
5. BASF (1958)

Appendix 9.   Carcinogenicity of PVP

D. Other Species

Species	Route	Duration	M.W.	Observations
Mouse (1)	24 subcutaneous injection of PVP in mothers and offspring.	> 500 days	Not stated	Lower total incidence of tumours and higher number of tumour free old females in PVP-treated mice born of PVP-treated mothers. Reduced incidence of spontaneous mammary tumours in treated animals compared with controls, also longer induction period.
Rabbit (2) (13–19/grp)	0.4–3.0 g/kg i.v. in 5–20% 2–14 times at monthly intervals	21–89 months	K-60 K-30 K-20	2 tumours in PVP K-20 grp (both gonadal seminomas) 2 tumours in saline-treated control (1 gonadal seminoma and 1 adenocarcinoma). No tumours in K-30 or K-60 groups, but histological evidence of PVP storage in liver and spleen in most rabbits. Some animals had been exposed previously to other products (for skin toxicity testing).
Dog (3)	In diet 5% or more	1 yr	37,900 K-30	Two studies reported together. No toxicity attributable to PVP (see Table 6B). No evidence of carcinogenicity.
Beagle dog 2/sex/grp (4)	In diet 10% PVP + 0% SF 5% PVP + 5% SF 2% PVP + 8% SF 10% Solka-Floc (SF) (SF = cellulose)	2 yrs	37,900 K-30	No adverse toxicity reported throughout study (see Table 6B). Slight swelling of reticuloendothelial cells in lymph nodes of group receiving 10% PVP, and some evidence of this in 5% and 2% treatment groups to lesser extent. No evidence of carcinogenicity reported.

1. Stern et al (1956) cited by Hueper (1959)
2. BASF unpublished report
3. Shelanski (1958) and Wolven and Levenstein (1957) cited by Burnette (1962)
4. Princiotto et al (1954)

Index

(Also see Appendix for summaries of all studies.)

T - #0169 - 071024 - C0 - 234/156/13 - PB - 9780367450830 - Gloss Lamination